Issues in Higher Education

Series Editor: GUY NEAVE, International Association of Universities, Paris, France

Other titles in the series include

GOEDEGEBUURE et al.
Higher Education Policy: An International Comparative Perspective

NEAVE and VAN VUGHT
Government and Higher Education Relationships Across Three Continents: The Winds of Change

SALMI and VERSPOOR
Revitalizing Higher Education

YEE
East Asian Higher Education: Traditions and Transformations

DILL and SPORN
Emerging Patterns of Social Demand and University Reform: Through a Glass Darkly

MEEK et al.
The Mockers and Mocked? Comparative Perspectives on Differentiation, Convergence and Diversity in Higher Education

BENNICH-BJORKMAN
Organizing Innovative Research? The Inner Life of University Departments

HUISMAN et al.
Higher Education and the Nation State: The International Dimension of Higher Education

CLARK
Creating Entrepreneurial Universities: Organizational Pathways of Transformation.

GURI-ROSENBLIT
Distance and Campus Universities: Tensions and Interactions. A Comparative Study of Five Countries

TEICHLER and SADLAK
Higher Education Research: Its Relationship to Policy and Practice

TEASDALE and MA RHEA
Local Knowledge and Wisdom in Higher Education

TSCHANG and DELLA SENTA
Access to Knowledge: New Information Technology and the Emergence of the Virtual University

HIRSCH and WEBER
Challenges facing Higher Education at the Millennium

TOMUSK
The Open World and Closed Societies: Essays on Higher Education Policies "in Transition"

NEAVE et al.
The European Research University: An Historical Parenthesis?

MEEK and SUWANWELA
Higher Education, Research, and Knowledge in the Asia-Pacific Region

SÖRLIN and VESSURI
Knowledge Society vs. Knowledge Economy: Knowledge, Power, and Politics

SAGARIA
Women, Universities, and Change: Gender Equality in the European Union and the United States

SLANTCHEVA and LEVY
Private Higher Education in Post-Communist Europe: In Search of Legitimacy

ENDERS and DE WEERT
The Changing Face of Academic Life: Analytical and Comparative Perspectives

HARPUR
Innovation, Profit and the Common Good in Higher Education: The New Alchemy

The IAU

The International Association of Universities (IAU), founded in 1950, is a worldwide organization with member institutions in over 120 countries. It cooperates with a vast network of international, regional and national bodies. Its permanent Secretariat, the International Universities Bureau, is located at UNESCO, Paris, and provides a wide variety of services to Member Institutions and to the international higher education community at large.

Activities and Services

- IAU-UNESCO Information Centre on Higher Education
- International Information Networks
- Meetings and seminars
- Research and studies
- Promotion of academic mobility and cooperation
- Credential evaluation
- Consultancy
- Exchange of publications and materials

Publications

- International Handbook of Universities
- World List of Universities
- Issues in Higher Education (monographs)
- Higher Education Policy (quarterly)
- IAU Bulletin (bimonthly)

Knowledge Society vs. Knowledge Economy

Knowledge, Power, and Politics

Edited by

Sverker Sörlin
and
Hebe Vessuri

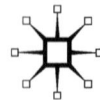

KNOWLEDGE SOCIETY VS. KNOWLEDGE ECONOMY
Copyright © Sverker Sörlin and Hebe Vessuri, 2007.
All rights reserved.

First published in hardcover in 2007 by
PALGRAVE MACMILLAN®
in the United States—a division of St. Martin's Press LLC,
175 Fifth Avenue, New York, NY 10010.

Where this book is distributed in the UK, Europe and the rest of the world, this is by Palgrave Macmillan, a division of Macmillan Publishers Limited, registered in England, company number 785998, of Houndmills, Basingstoke, Hampshire RG21 6XS.

Palgrave Macmillan is the global academic imprint of the above companies and has companies and representatives throughout the world.

Palgrave® and Macmillan® are registered trademarks in the United States, the United Kingdom, Europe and other countries.

ISBN: 978–0–230–11570–5

Library of Congress Cataloging-in-Publication Data
    Knowledge society vs. knowledge economy : knowledge, power, and politics / edited by Sverker Sörlin and Hebe Vessuri.
      p. cm.—(Issues in higher education)
    Includes bibliographical references and index.
    ISBN 978-1-4039-7304-7 (alk. paper)
      1. Knowledge, Sociology of. 2. Knowledge management—Economic aspects. 3. Education, Higher—Economic aspects. I. Sörlin, Sverker.
    II. Vessuri, Hebe M. C. III. Series: Issues in higher education (New York, N.Y.)
HM651.K564 2007
306.43'2—dc22   2006040597

A catalogue record of the book is available from the British Library.

Design by Newgen Imaging Systems (P) Ltd., Chennai, India.

First PALGRAVE MACMILLAN paperback edition: July 2011

10 9 8 7 6 5 4 3 2 1

Transferred to Digital Printing in 2011

Contents

List of Illustrations — vii

List of Acronyms — viii

Introduction: The Democratic Deficit of Knowledge Economies — 1
Sverker Sörlin and Hebe Vessuri

1. Modes of Knowledge and Patterns of Power — 35
   Maurice Kogan

2. Universities and Society: Whose Terms of Engagement? — 53
   Mala Singh

3. Knowledge, Globalization, and Hegemony: Production of Knowledge in the Twenty-First Century — 79
   Paul Tiyambe Zeleza

4. Knowledge and Equity: Unequal Access to Education, Academic Success and Employment Opportunities: The Gender Balance in Algeria — 107
   Nouria Benghabrit-Remaoun

5. Knowledge, Culture, and Politics: The Status of Women in the Arab World — 125
   Fahima Charaffedine

6. Knowledge, Theory, and Tension Between Local and Universal Knowledge — 137
   Roberto Fernández Retamar

7. The Hybridization of Knowledge: Science and Local Knowledge in Support of Sustainable Development — 157
   Hebe Vessuri

8. Remarks on the Relationship between Knowledge Functions
   and the Role of the University   175
   Akira Arimoto

Index   199

List of Illustrations

Tables

4.1	Lycée selection: Proportion of girls to boys	116
4.2	Educational success rates (%)	117
4.3	Distribution by type of activity and sex	119
4.4	Distribution of local paid employment by sector and sex, 1994	119
5.1	Structure of active women economically in professional, technical, administrative, and managerial posts in Arab countries (1970–1994)	130
8.1	Knowledge functions and their corresponding academic organization	178
8.2	Integration of research and teaching orientations	194

Figures

5.1	Average GEM (gender empowerment measure) values: World regions, 1995	128
5.2	Regional illiteracy rates by gender and gap (for early 1990)	130
7.1	Forms of knowledge	167
8.1	Knowledge, society, and university	180

List of Acronyms

AAU	(Association of African Universities)
ACU	(Association of Commonwealth Universities)
BEF	(Basic Education Certificate)
CODESRIA	(Council for the Development of Social Science Research in Africa)
COE	(centers of excellence)
COL	(centers of learning)
CUDOS	(communality, universalism, disinterestedness, organized skepticism, competition, and originality). *Indexed under* CUDOS
EZLN	(Zapatista Army of National Liberation Army)
FD	(faculty development)
FTAA	(Free Trade Area of the Americas)
GATS	(General Agreement on Trade in Services)
GEM	(gender empowerment measure)
ICTs	(information and communication technologies)
IDB	(International Development Bank)
IMF	(International Monetary Fund)
JUAA	(Japan University Accreditation Association)
KBS1	(knowledge based society 1)
KBS2	(knowledge based society 2)
MEXT	(Ministry of Education, Culture, Sports and Technology)
MFPE	(Ministry for Vocational Education and Employment)
NAFTA	(North American Free Trade Agreement)
NGOs	(non-governmental organizations)
NIAD	(National Institute for Academic Degree)
OECD	(Organization for Economic Co-operation and Development)
ONS	(Office of National Statistics)
UBE	(Universal Basic Education)
UNDP	(United Nations Development Programme)
UNESCO	(United Nations Educational, Scientific and Cultural Organization)
WB	(World Bank)
WTO	(World Trade Organization)

Introduction: The Democratic Deficit of Knowledge Economies

Sverker Sörlin and Hebe Vessuri

"Knowledge economy" and "knowledge society" are concepts that reflect the growing importance of knowledge in our contemporary world. They underscore that whether we speak of the economy, or indeed society as a whole, the knowledge component is so crucial that it can be used to characterize both concepts. Most people, not least academics, usually take satisfaction in this general statement.

After all, knowledge is almost universally considered to be a public good, and something that must be supported, although the arguments may vary from the emancipatory Hegelian or Habermasian to the normative Mertonian, the pragmatic state-centered Weberian, or Vannevar Bushian—and all the way through to the downright economical-instrumentalist Bernalian or even corporate-focused or Etzkowitzian—which have been articulated in science and education policy documents after World War II, and increasingly since the 1980s. Indeed, so cherished is knowledge that several combinations of these arguments can be used by the same government, organization, or individual. Knowledge rarely seems to be a problem, and it is almost always a good.

Perhaps it is precisely in this overwhelming support, both rhetorical and real, for knowledge that those tensions and contradictions can be sought that make up the central thrust of this book. If knowledge is as potent a source of social power as the concepts "knowledge economy" and "knowledge society" seem to suggest, we should certainly expect different interests to occur in the workings of *how* knowledge shapes societies. That is also the pattern that we see. Issues concerning knowledge are moving, on a broad front, from the specialized discourses of science policy and higher education to the center stage of modern politics of globalization and competitiveness. If knowledge is a key defining aspect of contemporary and future societies we should also expect knowledge to become an increasingly politically laden concept and one on which a range of social interests try and make a claim. This is also why we have experienced, in the past decade or more, an increasingly tense, divisive, yet rich and creative, debate on the conditions of knowledge production, dissemination, and absorption. It is no longer enough to leave these issues with the academic community. They are becoming increasingly important to larger groups. They are rapidly expanding beyond the OECD, their classical core nations, to embrace the better part of the worlds' 200 nations.

We claim that the two concepts imply radically different visions and ideals of the role of knowledge. To summarize: knowledge-based economies are growing all around us, but they do so without always acknowledging the democratic, ethical, and normative dimensions of science and scientific institutions. The knowledge economy is market-driven and performs according to a market ideology, which stands in a problematic but not necessarily conflicting relation to the norms and ideals of the knowledge society. The knowledge economies we live in suffer from a democratic deficit. This does not mean that they have to be overturned or rolled back—that opportunity may not even exist. But what seems clear is that the democratic deficit needs to be addressed if academic life and culture should survive in the era of fierce global competition, and if they should be able to spread and function in new regions of the world.

To describe and disentangle the inbuilt tensions that the two concepts imply is not just an academic enterprise, important and worthy in itself. It is a mission that will serve the needs for conceptual clarification and policy direction among communities worldwide that take an interest in the growth and management of higher education and science, and the particular social institutions where knowledge is at the core of everyday work and life. This means not just universities and colleges, although they occupy an important place in this growing universe of institutions. It also involves research institutes, corporate R&D labs and offices, departments of government on national and regional levels, and international organizations and foundations that make up the working and funding partners of the performers of higher education and research.

These topics were selected as the theme for debate of the first Global Research Seminar selected by the UNESCO Forum on Higher Education, Research and Knowledge in December 2003. In the context of higher education systems, structures, and knowledge production characterized by rapid global change, with an increasing gap between developed and developing countries, and in a political climate largely governed by neoliberal policies; the seminar was aimed to bring to the fore the following issues for critical analysis and debate: changing concepts of knowledge in the twenty-first century; conceptualizations of "knowledge society" and "knowledge economy"; relationship between knowledge and power: organization, actors, and structures; the role of the state in the context of knowledge production; impact of fragmentation of knowledge in the process of specialization; universality versus identity, language, and values; role of the intellectual and the future of ideas, thought, and critical analysis; hegemony of knowledge norms: alternative readings.

Worldwide, the landscape of higher education and the sectors of research and knowledge production are undergoing profound transformation. In the wake of liberalization of economies, the global mobility of citizens, capital resources, and knowledge, and the increasing demand of skilled labor, higher education systems and structures are under the immediate impact of changes driven by unprecedented global social and economic forces with the processes of change of higher education structures and of research thus being embedded in an extremely complex reality, in which no self-evident choices are available and where actions have multiple effects in a dynamically interdependent environment.

To assemble the full complexities, and the global variety, of these themes in one single book is, needless to say, an impossible task. Yet this is an attempt to provide at least an outline of the kinds of conflicts and tensions that are emerging in different regions of the world. Our empirical cases cover large regions and many states; already in the cases presented we can discern sufficient contrast and radical differences of conditions to be able to say that the coming several decades will be among the most interesting ever in the continued growth of higher education and science institutions. At the same time, we do identify common themes and threads that unite even the most contrasting institutions. It is the nature of knowledge, perhaps, to speak to any audience, any condition, and present challenges of deeply social and ecumenical nature that are shared from the most advanced science institutions to the least equipped and the needy. Indeed, in the words of J. Robert Oppenheimer, knowledge is in the deepest sense to do with "the power of betterment—that riddled word" (Oppenheimer, 1954, p. 97).

Utopias and Take-Offs

Early Utopian Notions—Industrialism and Imperialism as Obstacles

The tensions that this book addresses are deeply rooted. The ideas of a knowledge society and a knowledge-based economy, respectively, are not only speaking to different knowledge interests and knowledge communities. They also have their quite unique and different histories. Although neither became of common usage until the 1990s, when they did perform rapid and successful careers as concepts and thoughts, they antedate this decade with many years, if not centuries.

The idea of a knowledge society may in its oldest version be traced back to Bacon's New Atlantis and other knowledge utopias. Strangely, however, the growth of industrial societies stood in the way of more radical visions of knowledge societies well into the twentieth century. The idea was to some extent preserved in the utopian tradition. In the writings of Marx as well as Bellamy, Morris, and other visionaries in the late nineteenth and early twentieth centuries, the notion of knowledge and knowledge seeking came up as a component in the vision of future societies (Manuel, 1973). In that sense knowledge has been deeply embedded in the very idea of human emancipation, and increased productivity as an outcome of new technology has been at the roots of quite recent attempts at social reform such as the demand for a 6-hour work day, or French thinker André Gorz's utopian discussions of an almost jobless ideal society where 2 hours of "socially necessary" labor would be enough (Gorz, 1980)

These ideas, often rooted in the Marxian tradition, were very different from the growth of a more instrumental view of science as a tool of progress and enhanced competitiveness of firms and nations. Such ideas did not gain much prominence until well into the twentieth century. Some early aspirations were sparked off by efforts during World War I to create research councils and offices of technology (Leslie, 1993), and in particular after World War II, when the instrumental force of the combined science-military enterprise was demonstrated.

The slow growth of ideas of a knowledge society was probably a logical consequence of the basic tenets of the growth of new industrial technologies. These developed with only small and haphazard, if at all any, contacts with progress in science, and this was the case in most areas until the twentieth century. A future society, even an ideal society, could not realistically be defined by science as long as progress and economic prosperity were closely bound up with engineering, technology, and hardworking men and women in fields and factories.

Nor could knowledge societies be conceived of beyond the confines of the Western world. Despite the fact that there had been remarkable advances in science and technology outside Europe, both before the Renaissance in regions such as China and India in particular and later with the arrival of the Jesuits and other science-promoting agents from the West (Adas, 1989; Bayly, 2004; Kumar, 1995; MacLeod, 1982/1987), no realistic notion of autonomous scientific growth was possible in these parts of the world under the influence of imperialism. In fact, the very power structures of imperialism made the idea almost pernicious that knowledge societies, or anything like that, should occur in regions where even basic primary education was scarce and where scientific institutions were true rarities, and furthermore largely dependent on sustained European expertise, infrastructure, and political power. Exceptions were of course those areas of the world, colonized by large portions of European settlers who founded "neo-Europes" (Crosby, 1986) that socially, economically, and ecologically harked back to the Old World and were soon (beginning in the nineteenth century) able to emulate or even surpass it, even in science and higher education. This was true, to varying degrees, for Australia, New Zealand, South Africa, Canada, and the United States, where the pace of reception of anything from Darwin to Einstein was sometimes faster and more profound than in the Old World (Chambers, 1991; Reingold, 1966).

For other colonized regions, growth in educational and scientific institutions followed a much slower and poorer route. Only with the demise of imperialism in the decades immediately following World War II was it possible to discern, and start planning for, a growth of modern higher education and science institutions. This process, consequently, occurred in the newly born decolonized states. Exceptions were the states outside Europe that were never colonized or had been decolonized earlier: Japan, Latin America, parts of the Middle East. In these areas, universities and scientific institutes had in some cases, notably in Japan and Latin America, grown independently and occasionally gained considerable quality and reputation. On the whole their position was still far weaker than in Europe and North America, and ideas of knowledge societies could hardly gain any ground in those regions either. Japan was perhaps the nation where performance in science and technology came closest to Europe and the United States, but even there competitiveness in basic science remained modest even under its most successful period as an economic superpower in the 1970s and 1980s—although Nobel Prizes were high on the research policy wish list, and some were also received.

Take-Off for Knowledge As a Social Force—Post–World War II Definitions

As this broad-brush picture indicates, the two historical forces of industrialism and imperialism acted as effective obstacles to ideas of *knowledge* as a historically realistic defining feature of societies. The history of the idea of knowledge society, not to speak of a globalizing knowledge society, is, therefore, in practice a history, which could only occur in the post–World War II period. In the 1960s, Peter Drucker (1969, pp. 247–249) suggested that knowledge had indeed become the fundamental driver of modern societies; we were moving "from an economy of goods [to] . . . a knowledge economy." However, important elements of its thought structure were borrowed from earlier thought on science. It could use the normative structure of science (Merton, 1942). It could incorporate notions of the instrumental and societal uses of science proposed by J. D. Bernal in *The Social Functions of Science* (1939) and in his multivolume *Science in History* (1952–54). It could certainly also draw on C. P. Snow (1959) and the two cultures debate, and on Kuhnian ideas of the major structural changes in science, to argue that knowledge and science had to do with very basic patterns of social change.

Yet, a further component started to emerge only in the post–World War II era: the massification of education. Seminally observed and analyzed only in the early 1970s (Trow, 1970, 1974), the growth of mass higher education had started in earnest in North America (both Canada and the United States) in the 1950s, and Europe and other neo-Europes followed this trend. In essence, the knowledge society could only be conceived of as a society where a large portion, if not the majority, of the population had some elementary academic education, and where quite a few also had advanced degrees. This kind of society was not possible to envision until Berkeley sociologist Martin Trow published his influential articles on mass, and indeed "universal," higher education in the early 1970s.

At around the same time another equally influential study appeared, Harvard sociologist Daniel Bell's *The Coming of Post-Industrial Society* (1973). Trow's and Bell's works may be seen rather like mirror-image twins. Not only did they occur almost simultaneously, but they also reflected two major avenues along which modernization occurred. Bell's study made a strong point of the view, emerging from the growth theory of economics in the 1950s and 1960s, that the human capital factor played an increasingly important role in explaining the sum total of economic growth. The concept "human capital" had been presented by Jacob Mincer (1958), challenging and augmenting Robert Solow's (1956) classical article on the role of technology in economic growth, and gained some ground in the following decade to boom in the 1980s and 1990s (Blaug, 1976; Kiker, 1966; Laroche et al., 1999). However, it had, importantly, been taken up by the OECD already in the early 1970s and was used to launch science and technology as a staple ingredient in the growth recipes that the OECD recommended to its members, and for which the organization developed indicators and collected statistics from its members, based on the Frascati manual adopted in the Italian city by that name in 1963 (the eighth edition was published in 2002).

Bell used the idea of human capital, and in particular the growing evidence of increased productivity that it brought, which indeed was the basic explanation of the sustained growth in developed countries, to argue that the most advanced nations were moving toward a postindustrial society. This society was characterized by high rates of productivity, by the continuous urge by the workforce toward the nonindustrial sectors of the economy which were least prone to easy productivity gains. In particular, Bell's analysis led to the conclusion that as incomes increased, and basic needs (a concept advocated by psychologists and marketing specialists based on theories by Abraham Maslow (1954) and others) became saturated, consumers would turn larger and larger shares of their spending toward those sectors of the economy where productivity gains were harder, if not almost impossible, to attain: health, specialized services, entertainment, travel, the arts. Indirectly, this would lead to a continuously increasing demand for specialists to provide these advanced services, and the specialists were almost universally academically trained. This came on top of increased demands of scientifically trained staff in other sectors of the economy for production, transportation, bureaucracies, education, and so on.

Postindustrial society, as described, envisaged, and propagated by Bell, seemed to be the material underpinning of the mass growth of education of new generations that was presented/analyzed/advocated by Trow. Perhaps it was no coincidence that both Bell and Trow had been drawn from the sociology program at Columbia University in the 1950s where communications and the behavioral sciences thrived. In its embryonic version, knowledge society was a society where people behaved differently. It meant the emergence of the modern human being, the suburban college-trained consumer who kept the society of mass consumption going. This was also the era of Walt Rostow's ideas of *Stages of Economic Growth* (1960), envisioning a society of mass consumption.

Up to this point there was a certain confluence of ideas—economical, educational, behavioral, idealist—that helped form the early thinking of a new kind of society that had knowledge as a defining feature. But this was also the time where these influences started to diverge and separate. The Humboldtian principles, further elaborated by the Weberian and Mertonian counterparts, still reigned supreme in the university sectors of most countries until the 1970s, despite the wide differences between their respective university systems. This systemic and normative coherence would soon come under attack as the role of universities as engines of growth and agents of social change came on the agenda. The human capital school of economic growth, the concept of manpower planning—also gaining ground during the 1960s and 1970s—were joined later in the 1970s and in the 1980s by an emerging agenda of academic capitalism and managerialism, arguing for more efficiency and more social accountability in the university sector, in the understanding that this sector would continue to grow and become an ever larger recipient of public funds. These were still the early days in this emerging university ideal, and seminal and authoritative texts from the period, arguing for change in this direction are quite hard to come by. Much of the early groundwork was done in the United Kingdom, where drastic regimes of accountability were considered already in the early 1970s, and they were implemented under Margaret Thatcher's premiership in the 1980s.

In the 1980s, science policy documents in many countries emphasized economic benefits from science, along with the strategic and security dimensions that had been there all along since the 1940s. It was also more and more commonly accepted that the connection between research and economic and social development was a complicated one. The stubborn belief that economic growth came straight from R&D investment proved untenable when "stagflation" haunted the United States and "Euro sclerosis" crippled European economics during the 1970s and 1980s. Economic theory reinforced the doubts. Neoliberalism was, at least in principle, not as prone as Keynesian economics, nor indeed as neoclassical liberalism, to accept that it was the task of the state to step in and pick up the bill where market failures occurred and to bear the brunt of investments in R&D that private companies lacked the muscle to do themselves. The immediate result was stagnation in research funding and the first serious dip in the rising curve of research spending that had characterized OECD countries for 4 decades.

The United States was partly an exception. After a 1970s of caution and confusion the SDI, the Strategic Defence Initiative (or "Star Wars"), under Ronald Reagan's presidency, meant a drastic increase in science spending in areas such as information technology, space research, and applied mathematics. But it could be argued that the SDI was simply a match replay of the National Education Act of 1958 but applied to hi tech, not to the education system directly (Neave, 2004). Even in those areas came, ultimately, a drop in spending after the end of the cold war with the fall of the Berlin Wall in November 1989, when parts of the research that had been funded by the Department of Defence (DOD) were downsized or terminated altogether. California, which had persistently been the largest recipient of military research funds, went into a serious economic crisis. The University of California, with its militarized research budget, was particularly hard-hit, which led to major restructuring programs and an increase in the numbers of overseas, better-paying, students. It was already more than a decade ago that this university started talking of itself as "a university for the world, not for the state." Another of the very big projects, *the superconducting supercollider*, that had been on the drawing board in an advanced stage and planned to be built in Texas, was stopped by Congress. The reasons were mixed (fraud and opulence were in the picture) but at the heart of the concern was a lacking confidence in what big science could actually contribute to society and economic growth. This downturn hit, principally, basic research in the exact sciences, whereas applied and biosciences were less severely hit and could even thrive.

The Accountable University

Behind this new and harsher funding climate for science was not just a lack of confidence in the economic returns of big science. It was also a matter of how public spending should be organized, and whether spending should remain public at all. The main theme was that the state was too big and must be reduced, and those parts that were left with the state must be used more efficiently. This gospel came to be subsumed under the catchword *accountability*. It was perhaps most easily seen in the United Kingdom under

Thatcher, but the ideas spread both geographically and politically, and when the Labour Party assumed power in 1997 most accountability measures that had been introduced under the previous conservative governments were kept, and new ones were added, true with some new political emphasis, for example, on widening access to higher education.

In the 1990s there started to appear a range of seminal academic texts that in various ways analyzed the kind of change that was taking place. At last, there was an understanding presented that could encompass the seemingly reduced support for basic science and blue skies research, and, at the same time, the overall expansion of science and higher education. Changes in research funding also became part of the analysis. The rapid increase in competitive funding and funds for applied purposes and to improve national competitiveness were understood as instruments in the quest for accountability. There was an explosion of academic analyses of the connections between academic research and competitiveness in the early to mid-1990s. Equally, there was an emerging literature on the "new contract" of science with the state (Faulkner and Senker, 1995; Funtowitz and Ravetz, 1992; Gibbons et al., 1994; Ziman, 2000). Further analyses emerged from OECD and other agencies, maintaining that continued expansion of undergraduate enrollment was also closely related to economic growth. The OECD claimed that where a large portion of the workforce had an academic degree the rates of economic growth were persistently higher (OECD, 1996). This sat well with social and political ambitions in North European countries, particularly under left-wing governments, to widen access to higher education among lower social strata and underprivileged groups.

Ideals and Features of a Knowledge-Based Economy

It was from this line of thought, and from the economic interpretations of this literature, that the notions of a knowledge-based economy stemmed. Universities had always been important social institutions, but in a knowledge-based economy they were perceived as the prerequisites of the success of nations and, increasingly, of regions and cities. However, they would perform this function and merit their funds only if they were prepared to be flexible on their own norms and become more accountable to social and economic demands.

A central feature of this complex of ideas was that knowledge-based economies grew faster than other economies, even superfast. In the 1990s the U.S. economy grew considerably faster than in any other period since World War II. It did so without inflation and under low unemployment. This phenomenon was soon termed "the new economy." Economists were bewildered, even confused; historically the whole thing did indeed seem new, and with paradisiacal features.

When explanations were coming forth, however, they almost always included the increased role of knowledge. It could be the reaping of benefits from the IT and computer revolutions invested in previous decades, mostly in the form of productivity gains. Some would argue that the "knowledge content" of products and services had increased, and would regard that as a defining feature of the very notion of a knowledge-based economy. Productivity growth could be due to the effects of the surge in bioscience and biotech developments. And then again, it could be something else. Whatever it was, it must somehow reflect the fact that the U.S. population was now almost 50 percent college-educated. Moreover, the college-educated earned more and more money, both in absolute figures and in relative terms. The flip side of the 1990s economic miracle was that the American nation became even more divided than before. The number of low wage earners, often immigrants, grew enormously. The knowledge-based economy, may be a richer economy, but also a divided society.

The knowledge-based economy was to be regarded principally as a new phase of capitalism. To survive and thrive in the knowledge-based economy, it was now believed, the role of universities must change. The old norms and principles from Humboldt, Weber, and Merton would have to be compromised, if not discarded, and the principles of the "entrepreneurial university" (Slaughter and Leslie, 1997) would have to be put in its place where "third mission" activities are at the forefront and where patenting, license income, and cooperation with firms rank highly on the list of performance criteria. In short, the normative structure of science would have to be reformed and put in line with accountability and economically useful deliverables. Otherwise, the universities would fail to deliver on this increasingly central point. In this sense, knowledge economy purports to be a less ideological concept than does knowledge society. Nonetheless, it *is* ideological, at least in the context of university ideology and science policy. It serves to sustain an instrumentalist mission of the university.

Another feature of the knowledge-based economy points to the increased commodification of knowledge itself. This affects the inner workings of academe. Examples are abundant and can be gleaned from almost any corner of university life. Although there has always been a value in university degrees, the specifics of that economy were for a long time clouded behind a veil of ignorance. It was not evident to anyone what the real prize of a degree should be, or what the social investment of a degree should cost. Where there were college fees (for a long time only in the United States and Canada) the fees level was set quite randomly and probably too low to reflect market value. Not until the 1980s was the market for degrees being more formalized in the United States with widely publicized rankings guiding students, parents, and colleges as well in their quest to perform better academically and earn higher incomes. One effect of this has been a tremendous increase in fees in the past two decades. Among the indicators used is annual income a certain number of years after graduation, that is the market value of a degree influences the league position of a school or program.

Other features of the knowledge-based economy are found in the increasing international trade in degrees brought by student mobility. The countries most advanced in this aspect—for example, Australia, Canada, and the UK—bring large numbers of overseas students to their universities. In the case of Australia some 15 percent of the total higher education budget was derived from fees by foreign students' in the first years of the new century. The knowledge economy commodifies other forms of learning and teaching inside or outside academe: courses, programs, seminars. Evening programs grow. In the United Kingdom the Open University has continued its expansion and becomes increasingly global, a twenty-four/seven university for the world. Commercial online universities offer their services.

Similarly, in research, there is a trade with "star scientists" that are sold off and bought between universities and research institutes almost like football players. The phenomenon is still marginal outside the United States and the United Kingdom, but reflects the fact that grant-earning capacity, which is extremely unevenly distributed in the scientific community, is one of the key assets of a scientist as competitive funding becomes increasingly common. The star phenomenon seems unavoidable under current university trends, which means that steeper salary gradients will also become more common, just as is already the case in the United States, the market leader.

An Emerging Knowledge Society

The concept of "knowledge society," while sharing many of the same platforms in the early to mid-1970s, has taken a different direction. The economy has of course been a necessary component, and advocates of a knowledge society have been eager to demonstrate that if, or when, they get their way society will have more degree holders, more people working on and with knowledge, and more institutions producing and disseminating knowledge.

But there are other features that knowledge society advocates would add. They would first and foremost see this as an *ideal*, not just as a fact, which is perhaps the most basic distinction from the idea of a knowledge-based economy, which may, for some, be an ideal but would for most rather be regarded as principally a fact. The value dimension is clear and unequivocal. The question is whether society is good, fair, and just and if the knowledge society can improve it. Knowledge society proponents have put a greater emphasis on the public engagement in science, and on debate and discussion. Involving people in the scientific enterprise and a widening participation in higher education among all groups and strata of society has been among their goals.

Knowledge society, then, is a *historical* form of society, appearing at a certain time under certain conditions—almost like feudalism or capitalism. It is thus also a society that we may still not have attained fully but are approaching, in the same sense as one may be gradually coming closer to an ideal. Knowledge society, therefore, is an "epochal" concept, like agricultural society or industrial society. There is also an element of teleological thinking in it; sooner or later, knowledge society will come about. It is a good thing; it is an idealistic version of history. Nuggets of this way of thinking surface in Manuel Castell's major work on the "informational society" (3 vols., 1996–1998). Information society is emancipating people, releasing creative potentials and relieving hopes and fears. In the same way those who believe in "knowledge society" have described their future vision.

There is, as far as we are aware, no clear cut definition of knowledge society, other than the untold assumption that "knowledge" should be dominant, just as agriculture and industry have dominated earlier societies. Marc Uri Porat once defined "information society" as a society where a majority of the workforce performs information-related jobs (Porat, 1977). With a similar definition we should require a large portion, for example 50 percent, of jobs in knowledge society to be in or concerned with knowledge. The problem then is another definition: what is "knowledge"? In economic and sociological literature, the concept "knowledge sector" is sometimes used, but it does not solve the problem. Knowledge exists in so many versions, and in a certain sense knowledge is something that everybody uses. Therefore, advocates of knowledge society have to, pragmatically, resort to statistics of academically trained in the workforce.

Knowledge society has carried certain political overtones. It has been open to change, for rationality, for basic scientific values and methods. It has underlined the enriched life of the individual and the democratic gains for society at least as much as the economic benefits of increased knowledge.

Accountability and Normative Drift

Much of these differences between knowledge society and knowledge economy derive from the different connotations of "society" and "economy." The first concept speaks to the community, to the ties that bind groups, individuals, and institutions together into a whole that is larger than the sum total of the parts. The economy is a subcategory of society, a limited prerequisite. If you put the defining emphasis on the economy you tend to shy away from the communitarian aspects of knowledge, the meaning of it all, and concentrate on outputs.

From this conceptual analysis it is clear that the choice of concept, and the bundles of ideas and values that come with it, has an enormous influence on policy and, ultimately, on the way knowledge production and management is carried out.

Why should universities hold and carry certain core values? We are thinking here not only of autonomy as such a core value but also of the Mertonian CUDOS norms (Merton, 1942) and other rules of honor and indeed even academic liturgy and procedural order. The question seems all the more pressing if it is an established fact that performance and quality assurance are measured and managed through other institutions external to the individual university (which is the case because the university as an interested, publicly funded agent cannot be trusted with assessing its own performance, according to this logic).

In order to deal with this issue we must widen the perspective to include the full potential of the roles and uses of the universities in society. If one does that, one could think of two fairly distinctive approaches. The first has to do with *values as support to performance*. The second has to do with *values related to the wider social role of knowledge and knowledge-based authority*. The latter issue is too often neglected in discussions of managerialism, autonomy, and accountability.

However, is it a value issue, or an empirical issue? Although mission-oriented research groups seem to perform well, in fact very well, in basic research (Laredo, 2001), there is also overwhelming evidence to support the claim that basic research in traditional academic settings—research universities, with traditional academic funding and with a large degree of autonomy—is an enormously efficient way of producing new theoretical and empirical, if not immediately applicable, knowledge (Pavitt, 2004). If one was to answer the question of how a knowledge-producing system should be designed—as if on a clean slate—no single proposal would be a priori better than the other. Nevertheless, an answer that ruled out autonomy and CUDOS altogether would clearly be out of touch with everything we know about the scientific enterprise. The independent researcher, or research group, discovering truths about nature and society, driven by a quest to know, is not a piece of self-interested fiction, invented by the scientific community itself. It is a sound empirical fact.

This historical and sociological fact may be translated into contemporary policy debates simply by applying a set of empirical questions. What is the level of autonomy and CUDOS in different research environments—and how do these perform on a number of dimensions? This is a fully researchable question and it is already being turned into monitoring schemes in many countries. What remains, as yet, less well developed, is the relation between traditional research performance measures (publications, citations, level of funding, academic invitations, etc.)—that is, what might be called quality of research, on the one hand, and success in entrepreneurship on the other.

There seems to be some positive connection between a very high-level performance of research and business indicators—such as patents, firm start-ups, license income, and so on—although the literature is stronger on start-ups and firm formation than on licensing and patents (e.g., see Etzkowitz, 2002, on MIT; Lowen, 1997 on Stanford; and Saxenian, 1994).

In the United States, universities hold a strong position in this respect (Lach and Schankerman, 2003). Among the top ten income earners from licenses in the United States are several high-rating research universities such as University of California, Columbia University, University of Wisconsin, and Stanford University (figures from 2002 according to the Association of University Technology Managers). On the other hand, there are many top-rated universities that do not have any significant license income. In addition, there are a few of the major license takers that are not highly rated research universities. It is largely a randomized game, with many attempts and very few "hits." In other parts of the world, the discrepancies, or random relationships, between academic quality and license income seem to be even more persistent, although the data are still scarce. At best, we may claim that good research is a necessary but not a sufficient prerequisite for success as an entrepreneurial university.

Similar questions posed to individual universities can be addressed to entire university systems. However, the connections at systems level are far more complex, and traditional indicators do not seem to work. Evidence again is contradictory. Very complex, pluralistic systems, such as the American, is a high performer in terms of innovation. There are studies to sustain the claim that this is due to properties of the U.S. system—such as its competitiveness and its pluralism with a large number of research-performing institutions (Henrekson and Rosenberg, 2000; Rosenberg, 2000). The cases of Finland, Canada, and the Netherlands, all strong scientifically, have recently enjoyed considerable success. There is, however, also evidence to suggest that small and not particularly strong university systems at times can function very well if they are well embedded in the private and public innovation structures (e.g., Italy). On the other hand, strongly university-focused systems with persistently very high ratings on the science indicators, such as Sweden, do not always enhance entrepreneurial performance (Henrekson and Rosenberg, 2000).

Again, data so far can tell us very little in terms of hard evidence in favor of any one particular policy solution. Knowledge in this area should be able to develop an empirical performance measurement of different university systems and thus monitor progress much better in the future than in the present. If so, the conditions for policy making will change.

In conclusion, on this point, although we know that academic values such as autonomy are not always necessary to achieve good performance ratings, we can certainly not say that they work against performance. Moreover, to try to reduce autonomy radically, and transform universities more generally along the lines of the business enterprise, is clearly going to hamper overall performance, even though exceptional success could be registered in individual cases, perhaps in particular in third-mission activities.

Values and Legitimacy

The entrepreneurial university is, however, only one of several dimensions of the third mission. There are other kinds of services, or "goods," that universities are openly or tacitly expected to deliver, and that they have provided, to a large extent, under the Humboldtian era. These tacit dimensions of extramural service may be criticism (a reasonably nonpartisan platform for informed opinion), credibility, reliability, special expertise, and, perhaps most importantly, trust. There may be others, but making the list very long creates the risk of overstretching the civil benefits from the university to an extent where most universities will fail to meet the standard. Hence, caution is appropriate. The examples given may be sufficient, at least for the purpose of this introduction.

These are certainly honorable properties. Are we not right in questioning whether it is just the idealized self-proclaimed version of universities—the vice-chancellor's annual address—that they represent? Would other social institutions in a similar, or even superior, way not be able to provide trust (banks or insurance companies?), expertise (consultants or hospitals?), criticism (media or artists?), and platforms for opinion (media or think tanks?)?

In principle—yes. In reality—no. Even with knowledge production being increasingly performed in such institutions, as the proponents of Mode 2 claim (Gibbons et al., 1994), there will still be enormous differences, first in the sheer comprehensiveness of knowledge amassed; second, in, precisely, autonomy. Banks and insurance companies cater for readily identified interests. They cannot be asked to provide the sort of independent opinion, criticism, or advice that we can rightly expect from universities or from academic intellectuals.

This extended role of universities is one they perform by their very nature of being universities—according to a certain system of values. The extended role requires a certain minimum of standards and probably also a certain minimum of intellectual critical mass. Low-rating universities and institutions with a bad reputation cannot expect to enjoy the same credibility as the top-ranking institutions. Interestingly, this seems to be true both in market-driven and in publicly funded university systems. The general level of trust in universities exceeds that of all other institutions in most countries where such things are measured (Boer, 2002). We may conclude that the legitimacy of universities is closely tied up with how far they are able to serve as upholders of criticism, credibility, and trust. If such properties are dependent on autonomy and nonpartisanship—which they are—autonomy is indeed something to cherish and to foster.

We have now further underlined the dilemma described above. There may be cases where research institutions can do well and provide impressive performance and much value for money with less autonomy; such institutions would rate high on accountability. There may equally evidently be circumstances under which such institutions should rightly enjoy public funding. At the same time, if the large majority of the research and higher education institutions we call universities lose their autonomy in order to achieve that particular kind of performance and efficiency, there would likely occur a loss of legitimacy that would in all probability ensure and constitute a more serious threat to the university system than the gain derived from (potentially) improved (short- to middle-term) accountability. The autonomy-accountability tension is seriously aggravated once the dimensions of legitimacy and the wider social role of universities are taken into account.

New public management (Kjaer, 2004) wizards care little about trust, credibility, or criticism. We should expect this omission on the part of managerialists to follow from a very common disease among institutional consultants and experts: they fail to take into account the tacit, or we may say civic, properties of institutions.

This is not to say, of course, that cases for managerialism could not be made. Universities can obviously shape up their performance through all sorts of incentive programs, through competitive funding systems, and other devices. This has happened in the United Kingdom, where the Research Assessment Exercises (RAE) show an admirable record of improved research performance on the ground, and a largely healthy, although occasionally worrying, shake out of underperforming institutions.

Is all then good and well? Efficiency is enhanced whilst managerialism stays away from areas where it would clearly be dangerous and almost surely would do harm? To address this question, we also have to ask the following: what if tacit values and properties were to be harmed through the very workings of managerialist culture?

That risk cannot be eliminated. It remains a memento to all practice and study of policy in this area. There is, however, relatively strong resistance and continuity in the academic culture. Despite the policy logic that drives most systems toward a managerialist ethos (an ethos that is not just superimposed on universities by governments—universities do also adopt it deliberately in what they believe to be their own self-interest), there is just a modest to low level of managerialist response. There is always certain inertia in organizations and some deep-seated love of identity and tradition, not to speak of university "brand names." There is a virtue of being credited with trust and credibility in an era when precisely such intangible but cherished virtues are being eroded.

What the future holds is not known. We would predict that the changes in universities will be on a grand scale, but remains limited. Variety will probably increase and institutions will take different strategies forward, encouraged by the increasing exposure to external pressures of demands both market- and politically driven. This process is well worth studying. It contains enormous possibilities, alongside with the dangers, and we may, in the process, see new hybrids of institutions, the properties of which we do not yet know.

However, some of the research institutions that will likely thrive in the new circumstances we already know quite well. They include special and professional schools, tailoring their education and combining it with profiled research; research institutes, specializing in advanced and mission-oriented research, often in flexible alliances with firms and universities; research companies that can deliver quickly and that listen to the need of the customer; specialized colleges, focusing on the right niches and building solid research competencies.

Some of these organizations may score high academically, but this is not always their primary role. Some may also enjoy trust and credibility. Most of these institutions will not find trust the only, or even the most valid, criterion to measure their success. It is extremely important that in trying to understand managerialism, we also discuss these kinds of research organizations. In these organizations, so it appears, managerialism (although rarely called by that name) is the order of the day, rather than the problematic occasional intervention.

Next come the universities. They are not all similar. Some may even repeat properties common among the less comprehensive performers (colleges, institutes, special schools). However, by and large, they are a different category. It is reasonable to believe that universities, in particular the full-fledged comprehensive research universities, will react differently to the new logic of policy. In a European setting, this is an interesting pattern to follow. In particular, if a funding level of some magnitude were to emerge above the level of the nation-state, there will certainly be universities trying to achieve at the highest level and claim their funding from it. Others will naturally have to opt for other levels.

Lastly, if we return to the caricature dualism of ancients and modernizers pitted against each other, it is indeed a false picture. Nevertheless, we should perhaps take one or two steps further in the analysis and create the Hegelian *Aufhebung*, which is to see that both skeptics and enthusiasts are framed within a larger pattern of inevitable change. Within such a pattern of change, there are probably many more positions to take, along a continuum, than the dualist caricature would suggest. Maybe there is also room at this point for a timely articulation of a third way in policy, combining the *longue durée* values of autonomy and CUDOS to defend the legitimacy of the university as an institution. If so, it should pay due acknowledgment to accountability and performance monitoring and other democratic and ordering features. This last qualification is added to sustain and enhance quality, to secure public engagement with, and support for, an institution that can continue to deliver social value precisely because it is an institution of credibility, criticism, and trust.

The Expansion to the Wider World

Are tensions between knowledge society and knowledge economy important enough to merit a book? As we have seen, these concepts, however crude and lacking precise definitions—perhaps best viewed as ideal types in Max Weber's sense—summarize the kind of choice that faces not only policy makers but also academic leadership in their work to carve out the role of knowledge institutions in the future. We certainly do not believe that the choice involves a clear-cut either-or. Most nations will need and will have elements of both. But even more acute is this choice at a time when the knowledge-handling institutions are growing in size and in numbers in a way hitherto unprecedented.

The twentieth century could be called the century of the university. Although most of the prestigious institutions were founded in earlier centuries, some already in the Middle Ages, the size of the academic enterprise grew tremendously in the previous century. Compared with 1900 the figures in 2000 simply exploded, in fact at an increasing speed in the latter (1980s and 1990s) decades of the century. Global enrollment of students has multiplied, as has the number of PhDs, the number of institutions, the number of scientific journals, not to speak of the number of scientists and academic staff. Accordingly, the pace with which university budgets have grown is equally impressive. At the beginning of the twenty-first century, the richest American universities have endowments that dwarf the GDPs of developing countries. As a share of GDP, the OECD average of funds allocated to R&D alone, not including higher education, has gone from 1 to 2.5 percent in the last quarter of the twentieth century. The increase is as rapid in industry as in universities and other research organizations.

Still, this expansion has until recently taken place within a fairly limited part of the world. Of the resources spent on science and universities at the turn of the twenty-first century, more than 80 percent were spent within the OECD area. In the coming century this is likely to change dramatically, not just because the non-OECD world is much larger, with some 85 percent of the world's population, but because most projections hold that economic growth in these regions will make it possible for many countries to make sustained investment in higher education and science. How this will play out in detail is hard to foresee. Different scenarios are possible. But even if we stick to current trends and limit our speculations to a 30–40-year period, it is obvious that the growth will be huge.

The growth of enrollments and institutions in the developing world is exponential, despite the fact that it is far too small to meet the needs. In China alone, the number of students in 2005 exceeds 11.4 million rising to more than 31 million by 2020 (Zuoxu and Rongtan, 2005), despite the fact that hundreds of thousands study abroad every year. The annual output of science students and engineers is already on par with the United States. China spent US$85 billion on R&D in 2003 (up 3 times since 1998), which puts her at an impressive third position after the United States and Japan. Total R&D investment came to 1.3 percent of China's GDP. Both China and India have potentials to become scientific superpowers and have already reached the top 20 lists of total citations in some fields (such as Physics; http://www.in-cites.com/countries/top20phy.html), admittedly on a very large population base and still with fairly low citations per paper, which indicates a large article production with a rather small but growing impact. Africa, at the other end of the scale, does not reach 1 percent of total article production (OECD Science and Technology Indicators, 2004).

It is not possible to evaluate this development in any simple straightforward way. Science and technology have brought problems and increased risks to societies and to nature. It is also true that science has contributed to a modernization that has created friction and anxiety. It is no less true, however, that science and knowledge have contributed enormously to human welfare and freedom. So far, we may speak of the last half century as a period of great progress in the name of education, science, and scholarship. In fact, no period in human history has seen a faster growth in these areas. In an increasing number of countries the probability is now better than 50 percent that a person born in 2004 will receive a university degree in her or his lifetime.

However, the greater part of this growth in higher education, science, and knowledge has occurred in a relatively small portion of the world, more or less limited to the OECD member states. Within OECD the absolute majority of all activities occur in North America and Europe, and if we expand this region to include the European Union with its new member states in Eastern and Central Europe the dominance becomes even more overwhelming. A few indicators may suffice to establish the asymmetrical relationship. North America and Europe together hold 95 percent of the world's doctoral degrees granted and continue to outstrip the rest of the world totally in the production of new PhDs by a rate of 10:1. Of the world's scientific articles they are accountable for 75 percent (OECD Science and Technology Indicators, 2004). The region is home to the great majority of the world's university faculty, and virtually all of the high quality institutions are located within the region, with a few Asian contenders showing up on the horizon.

A recent ranking of the world's 500 top universities (according to a set of conventional excellence criteria) by Shanghai Jiao Tong University saw an all-encompassing hegemony for Western, in particular American, institutions. Of the top 100 only a handful were Asian (Japanese) or Australian, and none was based in Latin America. In the entire African continent, with 1 billion people—more than Europe and North America together—only a couple of South African universities made it to the top 500 universities in the world. Sweden alone had 4 on the top 100 list, like Holland. Italy, Finland, Denmark, Norway, Switzerland, and France also had one or a few (http://ed.sjtu.edu.cn/ranking.htm; but see van Raan, 2005 for a methodological critique). However, there are southern countries that have improved their scientific standing rapidly over the past decade or two: Brazil, Mexico, South Korea, to some extent India, to mention a few.

The only dimension where the developing countries are anywhere near the performance figures of the average OECD country is in student enrollment. In some developing regions there has been a considerable growth in the number of higher education institutions and also in enrollment figures. It is rarely enough to justify the concept of "mass higher education"—coined in the early 1970s to describe the development in OECD—in countries such as India, South Africa, Iraq, Iran, Singapore, South Korea, and a number of Latin American states. Again, African states lag persistently behind, although individual universities have both grown and improved their record of teaching and research (Dakar, Makerere/Kampala, Dar-es-Salaam, and a few others). Research training in African universities is still very limited, with the institutions mentioned among the exceptions along with several South African universities.

For the world to meet the challenges of the future it is mandatory that this situation can change. Higher education and science must become more evenly distributed around the world if political tension is to be eased and possibilities for economic and social development is to be improved. The role of education and science in this process will not be argued here. It will be taken as a given, and a point of departure, that knowledge and skills are at least as important for the future of the developing world in this century as it used to be for the developed and industrialized countries in the past.

Models of Understanding the Nature of North-South Relations in Science

History goes a long way to explain the existing patterns. However, our understanding has become richer only in the past two decades. Still, in 1967 George Basalla published his seminal article in *Science*, "The Spread of

Western Science." Basalla identified three phases in the diffusion of scientific ideas, methods, and technologies from the "center" (Europe) to the "periphery" (the colonized world): (a) import of people, skills, and instruments; (b) establishment of colonial centers observatories, laboratories, museums, universities, etc. staffed by the colonizers but with some training of the colonized; (c) the independence and postcolonial phase when institutions were built and/or developed by the former colonies but in close relationships with centers in the colonial powers.

Basalla's model reflected the traditional linear and one-dimensional view of the growth of the Western presence overseas. However, in the following decades this view was challenged by more interactive models. In the 1980s and 1990s, it was argued that colonial centers in fact had been both earlier and stronger than Basalla had assumed, in particular in areas where the colonizers settled in large numbers: the Americas, South Africa, Australia, New Zealand, but also in pockets elsewhere (the Middle East, North Africa). These models were called "The Moving Metropolis" (MacLeod, 1982/1987) or "The Locality Model" (Chambers, 1991, 1993). What these models underscored was rather the constant interaction between centers in Europe and their counterparts in other regions of the world (Drayton, 2000; Grove, 1995; Kumar, 1995). For example, research in the history of colonial science could demonstrate that the reception of new scientific theories and ideas (Darwinism, Relativity Theory, etc.) was by no means always slower in the periphery. Centers rise and fall; they come and go. This is not to deny the existence of patterns of dominance or that the networks involved hierarchies, still less that science and technology in the colonies were organized to serve political, economic, and military power.

Colonial higher education was quite limited and grew considerably only in regions with a large colonization and particularly when they were no longer colonies. In South America, colleges and universities have existed from the sixteenth century. Harvard and other colleges in North America existed since the seventeenth century. These institutions mainly served the purpose of educating the colonizers, although in Latin America Jesuits and missionaries established colleges as a means to convert local populations to Christianity. Higher education in regions where colonization was scarce was a very different matter altogether. In large parts of Asia and in almost all of Africa south of the Sahara, no higher education was offered to local populations.

If gifted local representatives did go on to higher education they did so in European institutions. The first students from the Indian Raj appeared in Cambridge and Oxford around 1900. In the following decades, intellectuals from the colonies were trained in universities and capitals in France, England, Portugal, Spain, and Holland, depending on which colonial power they were the subjects of. Some colonial students also went to the United States and

Canada. Still, their numbers were very low and the penetration of higher education in colonial elites at the time of independence was in most countries next to zero. University-trained professionals among the majority black population in Mozambique at the time of independence in 1976 were so few that they could easily be seated in a normal classroom; there were virtually none at all in the Belgian Congo when it became independent in 1961.

Postcolonial Patterns

It is understandable, in this situation, that the postcolonial countries should seek to establish their own national higher education and research systems. Considerable progress has been made in some regions, less in others. No uniform pattern is discernible. In Singapore, the rate of higher education degrees is on par with some Western countries. Yet, there are countries in Africa where enrollment for higher education is still counted in fractions of a percent of a cohort. Some former Soviet republics and other Central Asian states demonstrate performance indicators similar to developing countries of the Southern Hemisphere. What is also worth noting is that although all states are now independent from colonial powers, old ties and networks still exist and new ones continue to emerge. Indeed, the relationships with foreign and overseas institutions and centers and even commercial suppliers of education and services are even stronger now than in colonial times. The globalization of knowledge marks postcolonial patterns of higher education and research as much as it does in Europe and North America.

It is by no means an easy or straightforward task to talk of the strengthening of North-South relations in this area. Networks and contacts are already strong. Institutional links exist. Still, there is always room for improvement. Education levels must rise, institutions must be strengthened, new institutions must be put in place, and research—now very limited—must grow. How is this to take place? Which role, if any, can be played by European and North American countries and institutions, outside of the roles entertained already today? Many questions wait for an answer. However, only some could be addressed in this book.

Where are We Heading?

Against this background the whats, hows, and whys in higher education and research become acutely important. These issues will define the world's future, as long as the decision-making elites continue to be formed and shaped by these institutions and their workings. Up until now, elites around the world have been by and large educated in established Western institutions, with both positive and negative results. They have been shaped by scientific paradigms and systems of thought developed in the European tradition. The "imperial curriculum" (Mangan, 1993) has formed generations of leaders in the developing world.

While Western centers of education will still maintain an attraction vis-à-vis the rest of the world, at least for a considerable time, it is no longer true that elites will always emerge from those institutions. In some cases, we should expect local, regional, religious, or other systems of thought (if not of scientific practice), which will markedly depart from what we have understood as the scientific enterprise and the system of values (CUDOS norms etc.) we have hitherto expected to find in it. However, we do not believe that this is the main issue, although the threats of religious fundamentalism and ideological zeal should not be disregarded.

Far more central to the future of the knowledge enterprise, we believe, are the kinds of goals, values, and purposes that will guide knowledge-producing institutions. It is in that respect that we maintain that the ideal types of "knowledge society" and "knowledge economy" have something to contribute to our understanding. Current trends seem to suggest, quite understandably, that issues of economic utility and individual career making exert enormous influence on newly opened institutions of higher education in the developing world. This should come as no surprise. Indeed, the same could be said for the historical development of institutions in Europe and the United States. The ideology and self-understanding of the research university is a comparatively late innovation in the nineteenth and twentieth centuries. It has been key to the success of science and to the enormous growth in power, both economically and militarily, in these parts of the world in recent history.

How far will it be possible for higher education and science to steer under national legislations? The basic tenets of universities took shape before the modern state. In the past two centuries, these very internationally oriented institutions have been made part and parcel of national systems of education and recently of industrial and innovation policies, then of course also including other research-performing institutions such as government laboratories and specialized institutes. Whether this will happen in any similar way in developing countries remains to be seen. So far, the state has proven indispensable for planning and funding of higher education in most countries. However, with rates of economic growth picking up, a significant portion of private institutions have established themselves in these regions. With education made part of WTO/GATTS, states now also have to adapt to privatization and trade in degrees. This is clearly perceived by some as an advantage, at least when governments lack the funds to build institutions themselves. But strong concerns are being voiced that governments will loose control and that democratic guidance of higher education for national goals will become increasingly difficult (see the contributions in AAU, 2004).

The Book—Perspectives and Contributions

The values that will direct higher education and research under such circumstances have yet to be defined. Perhaps we could talk of this prospect as an example of a knowledge economy rather than a step toward a democratic knowledge society. When free trade in higher education is broached up in this book, a great deal of concern comes with it. By and large, the authors agree that higher education has multiple purposes and ends, not all reducible to narrow corporate understandings of the "knowledge-based economy." The only notion of engagement that makes sense in the complex landscape of higher education in different parts of the world is a multidimensional one whose internal tensions and often unpredictable consequences require deft steering and constant negotiation.

We may wonder whether such a multidimensional view of social engagement is sustainable for universities in poor countries in the developing world, universities acutely constrained in their engagement choices by local socioeconomic and political impediments, and by the disadvantaged positioning of their countries and regions within the global asymmetries of power. All this makes them more vulnerable to certain types of "reformist" discourses that relate to economic liberalization. Social engagement must encompass and advance values and goals that relate to the many dimensions of human development. For this to happen, the terms of the knowledge society will have to unshackle themselves from the monopolistic demands of the market and be reconceptualized to include political, social, and ethical considerations that currently are absent, only weakly gestured to or hinted at.

Just as institutions of higher education and research spread from Europe in the past, the diffusion of managerialism that occurs today also comes from the West. It is important to see the logic of managerialism in order to fully assess its ramifications and its usefulness in the developing world. As Maurice Kogan states, in a largely theoretical chapter, there is always a relationship between knowledge and power. It is not only the old Baconian dictum that "knowledge is power," but also that knowledge can seek alliances with power and vice versa. Kogan identifies a tendency by the state—at least in the United Kingdom, but implies that it is a Western phenomenon—to try and gain control of knowledge-producing institutions through incentives, evaluations, and other demands of accountability. This is no longer confined to the Western world, but it is a way of managing science and higher education that gains ground everywhere.

The cosmopolitan and internationalist role of higher education requires that the approach of engagement is global as much as local, combining moral and intellectual concerns about the nature of the emerging global society and its differential impact on local communities in different countries. Mala Singh asks what could be the content of engagement for universities in sub-Saharan Africa in countries whose populations, viewed on a global scale, are the most impoverished and threatened by the lack of the basic necessities of human survival and dignity. The enormous scale of human struggle against poverty, disease, drought, famine, civil war, political authoritarianism, and decades of debilitating structural adjustment programs provides obvious ground for social engagement by universities that are considered to be infrastructural sources of knowledge, information, expertise, agency, and activism, no matter how meager or impoverished their own condition. But what are the real possibilities to develop and sustain appropriate policies and practices for engagement, which are not taken over by exclusively entrepreneurial rationales, to become the driving hand in situations of extreme constraint in funding and resources?

Taking into account some of the key lessons of history with regard to social engagement and commitment, Mala Singh raises the issue of how universities in sub-Saharan Africa can make sense of the current imperatives of accountability and societal responsiveness in a setting of wretched living conditions, few opportunities for vast numbers of the population, weakened role of the nation-state and power of the market in shaping social development? Given the low levels of participation in higher education, and limited opportunities for higher education study in the face of enormous demand, universities will exacerbate "imbalances of socioeconomic class, gender and regional origin in the student population" unless they are able to insert mediating mechanisms to open up access, for example, "affirmative action" for underrepresented student constituencies (including the disabled), loan and scholarship programs, "external degree centers and distance learning initiatives" to reach students in more remote parts of the country.

> What possibilities of normative and strategic choice for engagement are there for universities that have themselves been devastated by funding neglect due to sharp reductions in state and donor support, "exploding enrolments" which put a strain on physical infrastructure like classrooms and residences, teaching capacity and quality of provision, overcrowded libraries with dated holdings and general deterioration of working conditions and demoralisation of staff all round? (Sawyerr, 2002, pp. 23–24, quoted by Singh, this volume)

It is right to question what impact the knowledge society discourse will have upon engagement initiatives in universities in sub-Saharan countries. The absence, paucity, and unreliability of public funds make the university more vulnerable to market- and donor-driven imperatives and less able to set and follow a coherent and multidimensional engagement agenda for effective education, research productivity, and societal outreach. The global frameworks and arrangements developed by the World Bank (WB) and the World Trade Organization (WTO) take on implications still more worrying for universities in poor countries unable to mediate the conditionalities of globally lending organizations or international trade agreements as they bear down on national policy. The influence of such frameworks in directing the content of engagement along narrow one-dimensional conceptualizations of the knowledge society could be disastrous indeed.

Paul Tiyambe Zeleza connects to this risk in a critical discussion of the way African institutions for higher education have developed in recent years. To a large extent, his critique echoes the voices of protest against managerialism that have been heard from inside academia in both Europe and North America. He does not stop there, but further argues that globalization itself is the problem. Globalization is the engine that drives the demise of norms and institutions. The process is global, but its manifestations are more sinister in Africa where institutions are new, vulnerable, and lack the resilience that comes with age, tradition, and resources. It is also a process that reflects the weakness of the state in many African countries. Weak and poor states have, understandably, little to put up to safeguard them against the lure of commercial higher education or external institutions setting up shop, providing to their citizens that which they rightly deserve but which the state fails to deliver. The implications are profound, and in his view almost completely negative. It is, he asserts, a "neoliberal onslaught on the entire fabric of the postcolonial model of development, a model in which the university occupied a central, multifaceted role."

Dimensions in the Democratic Deficit

Democratic principles are linked to equity, both as a value and as a pedagogical practice. In the developing world, inequity and a democratic deficit seem to be common currency. The school and training systems are real catalysts of social change, and obviously they could challenge those preestablished orders that place women at a disadvantage in the Arab world. The issue of gender in the Arab countries is very complex and deep-seated as it touches basic cultural values. Even in cases like Algeria, where despite the secondary role assigned to women in society, the education system is characterized by a predominance of girls over boys; nevertheless the world of work remains largely closed to them. Compared with the 74.2 activity rate for the male population in 1992, the figure for women was virtually marginal at 9 percent. Gender equality is accepted, Nouria Benghabrit-Remaoun tells us in her paper. But, equality in employment is not even though girls' academic success and certificates of qualification should enable them to have access to gainful employment without any discrimination.

If gender appears as the most notable wastage, the problem lies in the education system itself. There is a gulf between life and school. Deep discontent and frustration with the promises of social advancement through educational qualification, which the state made during the wealthy days of the oil boom, have left young people bereft of any serious educational or training plans in mind. In times of uncertainty, religion becomes once more of an active ingredient of identity. Here again, while fundamentalism has been on the rise, women and girls have become the preferred targets of those who think that because of their educational success, women have usurped roles that were not traditionally theirs.

Fahima Charaffedine deals with similar problems whilst emphasizing cultural values, political empowerment, and participation. She is concerned with the social structure in the Arab world. Her departure point is the first *Arab Human Development Report*. It highlights lack of freedom as a dominant factor in political life, describes the relationships between governors and governed, their impact upon creativity and invention, as well as on the marginalization and exclusion of different social groups. The legislative structure in Arab societies is also described as paralyzed by law-inhibiting procedures and inadequate in its legal conditions to protect different individuals and groups. The prevailing system of cultural values is the greatest cause, in her view, for the low position of Arab women, which is reflected in both the distribution of their responsibilities within the family and in the distribution of power within the political domain. Illiteracy, fertility, and lack of economic participation interact organically with culture and politics. The role of women is identified only with her capacity to bear children (the reproductive role), within a culture of gender discrimination based on a traditional and religious education system. Improving the status of the Arab woman in society seems to require objective conditions related to knowledge, jobs, and empowerment, as well as to other factors, such as the building of democratic political life, and a democratic social culture open to modern knowledge and culture.

The notions of knowledge economy and knowledge society implicitly assume different understandings of the way the world is organized. Clearly, the authors in this book, belonging to the social sciences and the humanities, are inclined more to think of the aims and failings of the knowledge society conceived as an ideal. Thus Roberto Fernández Retamar reworks the topic of asymmetry in the world examined through the dichotomy between "universal," which he identifies as "underdeveloping" (i.e., causing underdevelopment) countries and the "local," which he recognizes as the third world underdeveloped countries. He reminds us of the persistent discrimination by which thought, intellectual capabilities, and rationality have been seen as a patrimony of the West, and the instinctual and the irrational as the habitual place of the inhabitants of the developing world. That this confrontation between metropolitan discourse and the utopian projects of an autonomous society remains unsolved causes it to reappear from time to time in various guises, new and old.

Some of the questions haunting the non-Western world were succinctly expressed by Gabriel García Márquez when he received the Nobel Prize for Literature in 1982; Fernández Retamar raises these questions once more: "Why is the originality so readily granted us [the third world] in literature so mistrustfully denied us in our difficult attempts at social change? Why think that the social justice sought by progressive Europeans for their own countries cannot also be a goal for Latin America, with different methods for dissimilar conditions?"

The strength and resilience of these cultural values is reflected in the science establishment as it spread across the face of the world. Scientific institutions have been built in the image of those in the West, though often they have been pale imitations of them. Hebe Vessuri argues that often, independent of local conditions, the establishment of national research and development (R&D) systems in the Latin American countries, linked to the growth of the middle classes, blindly mirrored the principles and postulates in the developed countries and were closely associated with them, mainly through scientific research. This configuration, she insists, lies at the origin of the failure of R&D systems to solve some of their societies' most pressing problems. It helps explain why the needs of the majorities were not translated into explicit demands to their R&D systems and were not generally regarded as topics for scientific research. The broad masses in Latin America and other parts of the developing world have been reduced to a negative example, a symbol of backwardness. To all intents and purposes, their creativity has been denied in most fields, technical or productive.

The combination of global, regional, and national crises and continuous waves of criticism and dissent that permeated layers of Latin American society since the 1990s was compounded by the emergence of new currents within academia. These new voices denouncing the failure of projects to integrate the poor into the national general advancement brought to an abrupt end many of the illusions of the middle classes that, threatened with unemployment, a fall in both income and standard of living, are themselves haunted by the possibility of falling into poverty. The disarray that followed the collapse of developmentalist programs and its replacement by a facile adoption of the rhetoric of a global order has only deepened the problems.

The way out of the impasse in which these societies find themselves does not seem to lie in reestablishing privileges for minorities as in adopting economic, social, and cultural policies that tackle head-on the most pressing problems of the bulk of the population. This implies that national R&D systems in developing countries recognize the existence of (a) technological bases that do not necessarily coincide with standard ideas; (b) a broad labor base that includes so-called unskilled (which does not necessarily mean "without knowledge") sectors of society; and (c) a revitalized interaction between technology and culture. In recent years there has been a noticeable change of attitude. While previously the main concern was with those social actors serving as "gatekeepers," who were responsible for opening the way to modern technologies for development, there is now greater recognition of the importance of taking local technological know-how into account.

Interpreting Change

Japan seems to have taken the route of the economy of knowledge on the assumption that an orientation to information and knowledge will be increasingly promoted throughout the world, and therefore national economic productivity must have a closer relationship to academic productivity. Hence competition must necessarily be encouraged and promoted. Akira Arimoto reconstructs the historical evolution of the cluster of the former imperial universities (*teikoku daigaku*) that stood at the top of a doubly stratified system of universities, colleges, and professional schools, as well as across the national, public, and private sectors. After World War II, although the former imperial universities were nominally classified as the one category of *daigaku* (university) together with other national and private universities, these artificially and intentionally stratified hierarchies were persistently and purposefully maintained for many years, until the present.

Recent government policies have set out to convert the previous practice from protecting all institutions on an equal basis to one of selective funding, encouraging competition among all institutions, to raising institutions' international competitiveness, and to reinforcing their ability to participate in the increasingly open, worldwide market. In particular, policies now propose (a) to promote mergers and integration of institutions so as to reduce the present 100 national universities to a much smaller number; (b) to establish the national universities as independent institutions with a managerial structure; and (c) to build a group of 30 universities, which would be able to attain the highest international levels by introducing the principle of competition and involving a third-party evaluation system.

The effect of the market upon the university has been strengthened to a large extent, and a demand-supply mechanism was also introduced to encourage competitiveness among institutions. Government control and oversight vis-à-vis the university has also been strengthened, despite gestures toward deregulation and policies of "remote steering."

All in all, the pattern that the authors in this book sketch out is one of profound and rapid change, primarily in the direction of knowledge economies, but, here and there, also toward emerging knowledge societies, although, in the language adopted in this introduction, much less of the latter. What does this mean? How should it be interpreted?

For Immanuel Wallerstein (2001)—who also contributed to the UNESCO Global Forum Seminar out of which this book has grown—this change heralds a crisis in the knowledge system that reflects a more widespread crisis of the world system as a whole. The structures of knowledge have "entered a period of anarchy and bifurcation, just like the modern world system as a whole," he writes. The reasons for crisis in the world system are the simultaneously and globally rising costs for personnel remuneration, inputs of production, and taxation (due to the needs of welfare and social security provision). Universities met their cost crisis by increasing their market orientation, Wallerstein argues, and are therefore wearing away as intellectual institutions. He also opines that universities may cease to be a major locus of knowledge production and that profound changes may well occur in the disciplinary setup of future universities, particularly in the social sciences (Wallerstein, 2004).

This is, mostly, a tale of decline. As the corrective, he proposes the intellectual as a moral and political representative of the knowledge community. The intellectual, he suggests, is preferably a generalist and a social scientist, one who in our time is prepared to stand up and explain our ongoing "*structural crisis or age of transition.*" A reader may or may not agree with the full-scale version of Wallerstein's metahistorical analysis. Still, in the chapters of this book, one may see a solid rallying to his call. The chapters have been written by intellectuals in the best sense of the word. They analyze a major historical change—call it a crisis or a transition—in higher education and research. They do so not just by presenting facts and theory but in acting as ethical and political subjects in laying bare inequities and problems. These problems are global, as higher education is going global. Yet, they occur in each and every country, and all have their particular version or variety. Thus, the contributors to this book illustrate neatly indeed Maurice Kogan's dictum that all knowledge always has some species of relationship to the existing structures of power, though in this particular instance the relationship is less affirmative and mostly critical.

References and Works Consulted

AAU (2004) The Implications of WTO/GATS for Higher Education in Africa. In *Proceedings of Accra Workshop on GATS 27th-29th April 2004*. Accra: Association of African Universities.

Adas, Michael (1989) *Machines as the Measure of Men: Science, Technology, and Ideologies of Western Dominance*. Ithaca, New York, and London: Cornell University Press.

Basalla, George (1967) The Spread of Western Science. *Science* 156.

Bayly, C. A. (2004) *The Birth of the Modern World, 1789–1914*. Oxford: Blackwell.

Bell, Daniel (1973) *The Coming of Post-Industrial Society*. New York: Basic Books.

Bernal, John Desmond (1939) *The Social Function of Science*. London: George Routledge.

—— (1969) *Science in History (1952–54)*. 3rd ed. Harmondsworth: Penguin.

Blaug, M. (1976) The Empirical Status of Human Capital Theory: A Slightly Jaundiced Survey. *Journal of Economic Literature* 14(3).
Boer, Harry de (2002) Trust, the Essence of Governance? In A. Amaral, G. A. Jones, and B. Karseth (eds.) *Governing Higher Education: National Perspectives on Institutional Governance*. Dordrecht: Kluwer Academic Publishers.
Bok, Derek (2003) *Universities in the Marketplace: The Commercialization of Higher Education*. Princeton, NJ: Princeton University Press.
Castells, Manuel (1996) *The Rise of Network Society*. 3 vols. Oxford: Blackwell.
―――― (1997) *The Power of Identity*. Oxford: Blackwell.
―――― (1998) *End of Millennium*. Oxford: Blackwell.
Chambers, David Wade (1991) Does Distance Tyrannize Science? In R. W. Home and Sally Gregory Kohlstedt (eds.) *International Science and National Scientific Identity: Australia between England and America*, Australasian Studies in the History and Philosophy of Science, no. 9. Dordrecht: Kluwer Academic Publishers.
―――― (1993) Locality and Science: Myths of Centre and Periphery. In A. Lafuente, A. Elena, and M. L. Ortega (eds.) *Mundialización de la ciencia y cultura nacional* Madrid: Doce Calles.
Crosby, Alfred W. (1986) *Ecological Imperialism: The Biological Expansion of Europe, 900—1900*. Cambridge: Cambridge University Press.
Drayton, Richard (2000) *Nature's Government: Science, Imperial Britain and the "Improvement" of the World*. New Haven, CT and London: Yale University Press.
Drucker, Peter F. (1969) *The Age of Discontinuity*. London: Heineman.
Edgerton, David (2004) "The Linear Model" Did Not Exist: Reflections on the History and Historiography of Science and Research in Industry in the Twentieth Century. In K. Grandin, N. Wormbs, and S. Widmalm (eds.) *The Science-Industry Nexus: History, Policy, Implications*. Canton, MA: Science History Publications.
Etzkowitz, Henry (2002) *MIT and the Rise of Entrepreneurial Science*. London: Routledge.
Etzkowitz, Henry, and Loet Leydesdorff (1997) (eds.) *Universities and the Global Knowledge Economy: A Triple Helix of University-Industry-Government Relations*. London: Cassell.
Faulkner, Wendy, and Jacqueline Senker, with Lea Velho (1995) *Knowledge Frontiers: Public Sector Research and Industrial Innovation in Biotechnology, Engineering Ceramics, and Parallel Computing*. Oxford: Oxford University Press.
Funtovicz, Silvio, and Jerome Ravetz (1992) Three Types of Risk Assessment and the Emergence of Post-Normal Science. In S. Krimsky and D. Golding (eds.) *Social Theories of Risk*. Wesport, CT: Greenwood.
Gibbons, Michael, C. Limoges, H. Nowotny, S. Schwartzman, P. Scott, and M. Trow (1994) *The New Production of Knowledge: The Dynamics of Science and Research in Contemporary Society*. London: Thousand Oaks; and New Delhi: Sage.
Gorz, André (1980) *Adieux au prolétariat: Au delà du socialisme*. Paris: Galilée.
Grove, Richard (1995) *Green Imperialism: Colonial Expansion, Tropical Island Edens, and the Origins of Environmentalism, 1600–1860*. Cambridge and New York: Cambridge University Press.
Henrekson, Magnus, and Nathan Rosenberg (2000) *Akademiskt entreprenörskap: Universitet och näringsliv i samverkan*. Stockholm: SNS Förlag.
http://ed.sjtu.edu.cn/ranking.htm.
http://www.in-cites.com/countries/top20phy.html.
Kiker, B.F. (1966) The Historical Roots of the Concept of Human Capital. *Journal of Political Economy* 74(5).
Kjaer, A.M. (2004) *Governance*. Cambridge: Polity Press.
Kogan, Maurice, M. Bauer, I. Bleiklie, and M. Henkel (2000) *Transforming Higher Education: A Comparative Study*. London and Philadelphia: Jessica Kingsley.

Kumar, Deepak (1995) *Science and the Raj 1857–1905*. Delhi and Oxford: Oxford University Press.

Lach, L., and M. Schankerman (2003) Incentives and Invention in Universities. NBER Working Paper No. 9727, May, 2003.

Laredo, Philippe (2001) Benchmarking of RTD Policies in Europe: "Research Collectives" as an Entry Point for Renewed Comparative Analyses. *Science and Public Policy* 28.

Laroche, Mireille, Mérette Marcel, and G. C. Ruggeri (1999) On the Concept and Dimensions of Human Capital in a Knowledge-Based Economy Context. *Canadian Public Policy* 25(1).

Leslie, Stuart W. (1993) *The Cold War and American Science: The Military-Industrial-Academic Complex at Stanford and MIT*. New York: Columbia University Press.

Lowen, Rebecca S. (1997) *Creating the Cold War University: The Transformation of Stanford*. Berkeley, CA: University of California Press.

MacLeod, Roy (1982/1987) On Visiting the "Moving Metropolis": Reflections on the Architecture of Imperial Science. In Nathan Reingold and Mark Rothenberg (eds.) *Scientific Colonialism: A Cross-Cultural Comparison*. Washington, DC: Smithsonian Institution Press.

Mangan, James A. (1993) (ed.) *The Imperial Curriculum: Racial Images and Education in the British Colonial Experience*. London: Routledge.

Manuel, Frank E. (1973) (ed.) *Utopias and Utopian Thought*. London: Souvenir Press.

Maslow, Abraham (1954) *Motivation and Personality*. New York: Harper.

Merton Robert K. (1942) The Normative Structure of Science. In R. K. Merton (ed.) *The Sociology of Science: Theoretical and Empirical Investigations*. Chicago, IL: University of Chicago Press; new ed., 1973.

Mincer, Jacob (1958) Investment in Human Capital and Personal Income Distribution. *Journal of Political Economy* 66(4).

Neave, Guy (2004) Higher Education Policy as Orthodoxy: Being a Tale of Doxological Drift, Political Intention and Changing Circumstances. In A. Amaral, D. Dill, P. Teixiera, and B. Jongbloed (eds.) *Higher Education and the Market: Rhetoric or Reality?* Dordrecht, the Netherlands: Kluwer.

Nowotny, Helga, P. Scott, and M. Gibbons (2001) *Re-Thinking Science: Knowledge and the Public in an Age of Uncertainty*. Cambridge: Polity Press.

OECD (1996) *Technology, Productivity and Job Creation*. Document DSTI/IND/STP/ICCP (96)2. Directorate for Science, Technology and Industry. Paris: OECD.

OECD (2004) *Science and Technology Indicators*. Paris: OECD.

Oppenheimer, J. Robert (1954) *Science and the Common Understanding*. New York: Simon and Schuster.

Pavitt, Keith (2004) Changing Patterns of Usefulness of Industry Research: Opportunities and Dangers. In K. Grandin, N. Wormbs, and S. Widmalm (eds.) *The Science-Industry Nexus: History, Policy, Implications*. Canton, MA: Science History Publications.

Porat, Marc Uri (1977) *The Information Economy: Definition and Measurement*. Washington, DC: U.S. Department of Commerce, Office of Telecommunications.

Van Raan, M. J. (2005) Fatal Attraction: Conceptual and Methodological Problems in the Ranking of Universities by Bibliometric Methods. *Scientometrics* 62(1).

Reingold, Nathan (1966) (ed.) *Science in Nineteenth Century America: A Documentary History*. London, Melbourne, and Toronto: Macmillan.

Rosenberg, Nathan (2000) *Schumpeter and the Endogeneity of Technology: Some American Perspectives*. London: Routledge.

Rostow, W. W. (1960) *The Stages of Economic Growth: A Non-Communist Manifesto* Cambridge: Cambridge University Press.

Sawyerr, Aki (2002) *Challenges Facing African Universities*. Accra: Association of African Universities.

Saxenian, AnnaLee (1994) *Regional Advantage: Culture and Competition in Silicon Valley and Route 128.* Cambridge, MA: Harvard University Press.

Selznick, P. (1957) *Leadership in Administration: A Sociological Interpretation.* Harper and Row Publishers; new ed., Berkeley, CA: University of California Press, 1984.

Slaughter, Sheila, and Larry L. Leslie (1997) *Academic Capitalism: Politics, Policies and the Entrepreneurial University.* Baltimore: Johns Hopkins University Press.

Snow, C. P. [Charles Percy] (1959) *The Two Cultures and the Scientific Revolution.* Cambridge: Cambridge University Press.

Solow, Robert M. A. (1956) Contribution to the Theory of Growth. *Quarterly Journal of Economics* February.

Trow, Martin (1970) Reflections on the Transition from Mass to Universal Higher Education. *Daedalus* 90.

———— (1974) Problems in the Transition from Mass to Universal Higher Education. In *Policies for Higher Education.* Paris: OECD.

Wallerstein, Immanuel (2004) "Knowledge, Power and Politics: The Role of the Intellectual in an Age of Transition." UNESCO Global Forum on Higher Education and Knowledge Working Paper. Paris: UNESCO.

Ziman, John (2000) *Real Science.* Cambridge: Cambridge University Press.

Zuoxu, Xie, and Huang Rongtan (2005) Research on the Macro Regulation Model of China's Mainland Post Secondary Education Expansion. *Higher Education Policy* (IAU) 18(2), June.

CHAPTER ONE

MODES OF KNOWLEDGE AND PATTERNS OF POWER

Maurice Kogan

Introduction: The Argument

In social science we struggle to discover contingent relationships and, perhaps too often, confuse these with what are no more than partial and contestable associations. This chapter attempts to note the extent to which knowledge and power may affect each other whilst pointing out that those interactions are less of a determinant nature than some analysts and social practitioners assume.

There are many discussions of the ways in which knowledge is shaped according to the field or tasks to which it is directed; this chapter makes an attempt to pick up one derivable theme from these concerns. This chapter tries to identify the extent to which modes of knowledge can be associated with different patterns of and assumptions about power; in addition it discusses the meanings and scope of power, both within, as well as beyond, epistemic communities and its bases, for example belief in specialization and peer evaluation, as against "social robustness" (Nowotny et al., 2001), implying more democratic or inclusive forms of evaluation. It explores the range of knowledge modes and analyzes their links with forms of power—and attempts to establish the dynamics of those relationships and shows them to be "multimodal" rather than *simply contingent on each other*. From there on it ponders on particular examples drawn from governmental and policy practices.

The underlying argument is as follows. Specialist knowledge has *intramural* or *internalist* power. It is governed by accepted rules of certification within epistemic communities. Whitley (1984, pp. 25–29), for example, argues that "developments in scientific fields are driven by a shared concern of participants with the establishment and maintenance of their reputations, and hence that such fields can be described as reputational work organizations".

But its second level of power is secular and depends on the scientist being able to convince the nonscientist that the work is useful or interesting. The converse might be true. Knowledge that begins by appealing to the "shared meanings of given social communities" or "social robustness" might gain power with user groups and gain purchase within those who share its epistemic ideology, but then might need to demonstrate sufficient amounts of of the test and demonstration features of hard science for it to be accepted as part of the scientific, *intramural* system, and gain credibility outside the scientific power groups.

Meanings and Scope of Definitions of Power

Our starting point must be the meanings and scope of power itself, in order to see whether its defining characteristics entail any particular knowledge components or styles, or whether its implications for knowledge are more the result of operational or instrumental frames within which it is enacted. In doing so, it will be necessary to divest ourselves of some oversimple assumptions, many of which emerge as dualities representing apparently contingent relationships. Thus, to take obvious examples, which we will focus more on later, positivist forms of knowledge generation are held to be associated with determined and statist forms of government. That is not necessarily so. Again, knowledge is thought to be "power," but, in fact, it can also be "disempowering," as when academics in the social sciences bar themselves from policy involvements by adopting critical stances.

General Accounts of Power

For our purpose we need hardly differentiate power from authority, but note that they are closely connected; "the latter has a normative dimension, suggesting a kind of consent or authorization, about which the former is similarly agnostic" (Isaac, 2004, p. 57). The power of knowledge may indeed become authority, which we can take to be an institutional subset of power. One account refers to authority as "a distinctive form of compliance in social life" and offers three accounts of the basis of this special compliance: authoritative institutions "reflecting the common beliefs, values, traditions and practices of members of society"; political authority "offering a co-ordination solution to a Hobbesian state of nature, or a lack of shared values"; and a third view that argues that "although social order is imposed by force, it derives its permanence and stability through techniques of legitimating, ideology, hegemony, mobilization of bias, false consensus and so on, which secure the willing compliance of citizens through the manipulation of their beliefs" (Philp, 1985). These accounts, however, are less definitional than descriptive of the genesis and consequences of authority. Some of the broader definitions of power may be more useful.

Isaac's discussion of power records four models:

1. The *voluntarist* model. For Dahl (1968) "power is a capacity to get others to do what they otherwise would not do, to set things in motion and change the order of events." "Power terms in modern social science refer to subsets of relations among social units such that the behaviour of one or more units (the response units, R) depends in some circumstances on the behaviour of other units (the controlling units, C)."

   This may be disputed by Lukes (1974), Bachrach and Baratz (1970) but invites the point relevant to our discussion that some forms of power depend on persuasion. Persuasive forms of knowledge, depending on rhetorical strategies, are likely to be different in format and content from those that depend on coercion or sanctions (i.e., authority) for their acceptance. There are links between these persuasive forms of the *voluntarist* and the *hermeneutic* or *communicative* models.

2. The *hermeneutical* model of power holds that it is constituted by the shared meanings of given social communities.

   This definition can be related to the way in which academic power is exercised. Within epistemic communities and their bases, the dominant source of power is an emphasis on specialization and peer evaluation related to it, which justifies exclusiveness—the specialist possesses knowledge not available, or less available, to others. The exclusiveness bites on those not empowered by specialized knowledge, regardless of the fact that it is shared within the epistemic community. Those within the peer group gain power and authority by their participation in the knowledge. In that sense, power *is not* a shared meaning but an exclusive and esoteric one.

   This is a perspective powerfully backed by Bourdieu (1975). He argues "that even the purest science is a 'social field,' with its own distribution of power and its monopolies, struggles and strategies, interests and profits. The scientific field is the locus of a competitive struggle for the monopoly of scientific authority." The better resourced and the more autonomous the field, the more tightly drawn the group of people that determine the holding of authority: that is the key competitors in the field. He thus not only distances himself from the idealized notion of scientific community but also insists that "the operation of the scientific field itself produces and pre-supposes a specific form of interest." Recent attempts rely on "social robustness" (Nowotny et al., 2001) implying more democratic or inclusive forms of evaluation, although this is, perhaps, more a program for action than a statement of what now dominates the fields of knowledge. Much earlier, Trist's (1972) definition of domains, too, implied multiple reference groups. In contrast to disciplinary knowledge, socially robust forms may generate power by their appeal to wide constituencies including those holding power within client and practitioner groups. Lindblom (1990), too, has articulated the case for demotic forms of "probing," which would be set fair to demote the power of academic specialization.

3. A *structuralist* model is rooted in the work of Marx and Darwin. It insists on the pregiven reality of structural forms that both enable and constrain human conduct. This leans toward power being vested in those who have command of the structures controlling knowledge formation and use.

4. In a *postmodernist* mode, as developed in Foucault (1977) and some feminist writing, language and symbols are central to power. Power is defined as "the capacity to act possessed by social agents in virtue of the enduring relations in which they participate." "It has a 'materiality' deriving from its attachment to structural roles, resources, positions and relationships." This microanalysis of the power exercised by different communities implies that knowledge is an exercise of power, which could be particularly exemplified by the power of academic disciplines. These power/power attributes can all yield some linkages with knowledge. The hermeneutical model, the more democratic or inclusive forms of evaluation, such as Lindblom's (1990) probing, Trist's (1972) domains, imply that the power that they generate may come through persuasion and interaction and their perceived utility. Both the voluntarist and structuralist models give space to the pressure exerted on the exercise of knowledge preferences by social structures such as academic status hierarchies or collegia. But this opens up the question of what kinds of knowledge will be more persuasive within these inclusive interactions, a question that we will discuss later in the chapter.

The Spectrum of Knowledge

There are several accounts of differences between different disciplines and areas of knowledge (e.g., Becher, 1989; Biglan, 1973; Storer and Parsons, 1968; and Whitley, 1977). These are largely concerned with their internal characteristics and shaping, rather than factors that might affect their relationships to external power. For our purpose, we offer a spectrum of knowledge ranging from "hard" and rigorously defined states to "soft" forms that are less capable of meeting the criteria of being "explanations which are at once systematic and controllable by factual evidence" (Nagel, 1961):

The Spectrum of Knowledge

HARD			SOFT
Hard science	Experiential/ Connoisseurial	Hermeneutic Phenomenological	Common sense (Nagel) Ordinary Knowledge

Nagel sets the scene for the hard end of the spectrum: "The practice of scientific method is the persistent critique of arguments, in the light of tried canons for judging the reliability of the procedures by which evidential data are obtained, and for assessing the probative force of the evidence on which conclusions are based." Such "internalist" models of science (i.e., those relying on exclusive and intramural governing arrangements) have exerted a powerful influence not only on scientists but also on those who have admiringly observed the growth and strength of science.

In the "internalist" view, science is an authoritative and self-regulating universe. The nature of scientific work, its evaluative criteria, and its institutional norms and structures are regarded as logically connected and rooted in the relationship between science and the physical world. The goals of science are "the extension of certified knowledge" (Merton, 1957). Science uncovers regularities of nature through accurate observation and empirical testing. It expresses and explains them in laws that are both as precise and as general as possible. The criteria of scientific merit are thus accuracy of observation and measurement, replicability of experimental work entailing rigor in design and control, validity and systematic importance or profundity of theory. The derivative and tightly interconnected technical and moral norms of logical consistency, emotional neutrality, and impartiality are strongly embedded in the classic statement of Merton (1942, 1957) of the four sets of "institutional imperatives" of modern science: (i) universalism; (ii) communalism; (iii) disinterestedness; and (iv) organized skepticism, and in the additional norms, is identified by himself and others of, for example, originality, humility, and independence. In this list, it should be noted that universalism and communalism are credited with belonging to the intramural versions of scientific power.

At the other end of the spectrum, there is Nagel's (1961) account of "common sense," and Cohen and Lindblom's (1979) "ordinary knowledge." Yet, if the "softer" forms of knowledge do not display "the organization and classification of knowledge on the basis of explanatory principles" they may yet seek "to discover and to formulate in general terms the conditions under which events of various sorts occur, the statements of such determining conditions being the explanations of the corresponding happenings" (Nagel, 1961, p. 13). They appeal, however, as much to the demotic and lay perceptions of what applies and what works as to any esoteric form of knowledge structure. (This should not be taken to imply that nonparametric knowledge cannot be rigorous, elegant, and, for that matter, esoteric).

Biglan (1973) sets out to co-cross-correlate three subject matter dimensions (hard-soft, pure-applied, life-nonlife) with the degree of academics' social connectedness, their commitment to teaching, research and service, the nature and extent of their publications, and their graduate training activities. His concern is, however, primarily with intraacademic organization rather than with those properties that generate or respond to the external exercise of power.

Claims, to be organized and classificatory, could apply across the boundaries and could apply not only to the "softer" forms of knowledge but also to historical analyses of changing polities and economies. Historical studies have moved a long way from Fisher's admission (1935) that "[O]ne intellectual excitement has . . . been denied me. Men wiser than I have discerned in history a plot, a rhythm, a predetermined pattern. . . . I can see only one emergency following one another, as wave follows upon wave . . . there can be no generalizations, only one safe rule for the historian; that he should recognize in the development of human destinies the play of the contingent and the unforeseen."

Within the internalist model, Polanyi (1962) argued that the validity of scientists' work is enforced not by objective proof but by the exercise of responsible judgment. For Popper (1972) the power of science is rooted not in its outcomes but in its methods of putting its propositions to the test. The issues became further elaborated and the *internalist* perspective to some extent undermined by Kuhn's (1972) belief that paradigms challenging previous conceptions are determined not only cognitively but socially by disciplinary communities. Mulkay (1979) went further and argued that recognition by the profession is the key objective of scientists and that the scientific community was not a republic but a complex nexus of problem-focused, discipline-centered, and wider networks of elites able to perpetuate themselves through interaction among differential allocation of resources, differential capacity to recruit the best talent, and a privileged informal communication system. With Kuhn (1962) and Mulkay (1979), "power rests not solely on epistemics but also on social arrangements."

The context of the analysis is thus one in which the salience of the concept of power, and particularly the power of knowledge, has been questioned through the sociology of knowledge and postmodernism—itself an important example of the power of knowledge to change political relationships. Assumptions about both knowledge and power have shifted. Concepts of power have changed markedly since the 1960s when power and authority as exercised politically and socially faced a crisis of legitimacy. The "hard" versions of science sustain legitimacy through their claims to impenetrable specialization backed by peer review. These venerable legitimacies have not been supplanted so much as paralleled by new ones. Knowledge may be authorized as much by its social robustness and relevance as by its epistemic containedness. It can be derived from communicativeness that is central to the hermeneutic and experiential modes of knowledge, though some of those working in the hard sciences might question whether one can always have confidence in what is being communicated. Perhaps the knowledge that scores highest is that which is "hard and tight," perhaps positivistic and quantitative in the social sciences, and/or which is geared to key public issues and explained through multiple media—the science of DNA technology would be a good example.

The power generated by knowledge might thus be affected by three sets of characteristics: the first concerns its communicativeness and appeal to social utility. The second concerns who determines the objectives of enquiry, researchers or government or industry? The third gets down to the heart of our concerns, according to Whitley (1977) and Weingart (1977), in relating the epistemic style and status of research to its power.

On communicativeness, Rip (1997) observes that the authority of basic science is legitimized not only by being fundamental and subject to rigorous testing, but also by the promises made by it. Industry shares the scientific view that basic science will yield results, a view shared by some of those participating in the United Kingdom (UK) Foresight Initiative (Henkel et al., 2000) who maintained that the "Foresight Initiative" needed different forms of knowledge, both "hard" and "soft."

On the objectives-setting dimension in, for example, health policy making, it has been noted that "applied research might be more readily useable by a policy system than basic research, but policy-makers tend to relate more willingly to natural sciences than social sciences. Research that follows priorities determined by the researchers themselves, according to the '*internalist* norms of science' is more often, though not always, going to be basic. Applied research is more likely than basic research to be following an agenda driven by forces other than the scientific imperative. . . . [W]here such drivers and sponsors are also the most likely potential users of the research, this provides some of the circumstances that might encourage utilization" (Buxton et al., 2002, pp. ii–iii).

The relationships between the producers and users of research have been described as follows: "The underlying power relationships can be various. Some researchers work within a managerial hierarchy in which they are subordinate to policy-makers; those working within government departments are obvious examples. Others work within a market in which the knowledge is purchased on the basis of competition with other researchers. For the most part the relationship is that of a market in which exchange and negotiation are the styles adopted. In such cases knowledge is exchanged for resources and legitimacy. Some market arrangements, however, allow for quite substantial tenurial rights which weaken the pull of the market and emphasize the need for well-constructed negotiation and exchange" (OECD, 1995).

On the epistemic set of considerations, we look for the ways in which the knowledge content, in itself, affects its power or autonomy. Whitley's comparison (Whitley, 1977) of restricted and unrestricted or configurational science is important. It shows how the cognitive structures of different sciences give rise to different forms of organization and so to different degrees of cohesion and power. The arithmetical ideal and the aim of expressing theory inhibits challenge in restricted sciences such as physics, concerned with a small number of properties of objects that can be quantitatively related. The high degree of specialization needed creates clear boundaries within these sciences, bureaucratization in the organization of research, and success in attracting resources. Configurational sciences, such as social sciences concerned with small numbers of highly structured entities exhibiting a large number of properties, are essentially polyparadigmatic: their conceptual boundaries are "highly fluid and permeable. In consequence their organization is less structured and there is greater scope for dispute and fundamental challenge." Although Whitley (1984) *does not* make the point directly, it can be inferred that this in turn affects their power outside their boundaries.

So too does, the "finalization" thesis of Van den Daele et al. (1977) make up the link. It identifies three phases of discipline development: (a) the exploratory, pre or polyparadigmatic phase; (b) the phase of paradigm articulation; and (c) the postparadigmatic phase. In the first and third phases problem orientation and discipline development are compatible. But when work is beginning to crystallize on the development of key theoretical models, usually the research program is dictated by "internal" needs incompatible with "external" problems.

Adding to the epistemological debate about the most appropriate forms of production of knowledge intended for utilization Trist (1972) argued that domain-based research represented a third category alongside basic and applied research. Domain-based, or policy-oriented, research is essentially interdisciplinary, and the crossing of new boundaries and the creation of new syntheses may advance both knowledge and human betterment. It also entails wider reference groups, beyond the scientific or clinical communities. Along similar lines, Gibbons et al. (1994) claim to identify a shift from the traditional discipline-centered mode of knowledge production that they characterize as Mode 1, toward a broader conception of knowledge production described as Mode 2. In this, knowledge is generated in a context of application and addresses problems identified through continual negotiation between actors from a variety of settings. The results are communicated to those who have participated in their production. Although the degree of change described by Gibbons et al. (1994) could be exaggerated, this general approach, as with that of Trist (1972), is compatible with attempts to identify power through utilization by explaining research production in terms of the interests of, at least, some potential users.

Modes of Governmental Power

We should now consider whether particular knowledge modes denote or support particular modes or styles of public activity, policy, or government. To keep the argument simple, we refer to central government at the head of systems. The classic and idealized models of government assume that government has its own power and power relationships and regulatory, allocative, rewarding, and sanctioning functions. They refer to somewhat autonomous entities but essentially capable of going their own ways without interpenetration or significant mutual effect. The simplicity of these classic assumptions has been drastically undermined in the past 40 years. We accept that both science, and/or, more broadly in the term of Cronbach and Suppes (1969), disciplined enquiry, and government inhabit worlds and client groups. But increasingly they have been pulled into each others' orbits.

The extent to which governmental power is strengthened by its commissioning and use of research has been shown to vary according to the salience of the policy field, the nature of the subject discipline or area to be employed, and the extent to which government at any particular time is committed to a display of evident rationality (Kogan and Henkel, 1983). It also varies according to the nature of the receptor (Caplan, 1977; Kogan and Henkel, 1983, 2000). The determinant forms of knowledge, "explanations which are at once systematic and controllable by factual evidence; . . . the organization and classification of knowledge on the basis of explanatory principles . . ." (Nagel, 1961, p. 4) may be more convincing to managers and politicians seeking certainties than will be a "softer" and less-controlled form of evidence. Perhaps more inclusive forms of encounter lend speed to persuasion, but the content of the persuasive message can be "hard" or "soft."

Factors Affecting Nature of Power Patterns

Sponsorship

We can now turn to identify those elements of research initiation and control that create power patterns that might frame knowledge creation, and the extent to which sponsors set or influence the setting or objectives. First, there is the nature of the sponsorship. Some knowledge creation is free of external sponsorship, but this is increasingly unusual. Perhaps it may be said to exist in those subject areas, mainly the humanities and social sciences, where academics are on tenure and require no more than a good library and a computer to produce a solitary or even a group work. Some mathematicians and philosophers may require even less—a pencil, note paper, and a glass of water. In the sciences and technologies advancement of knowledge usually requires money for equipment, materials, and technical backup. And certain types of social science depend on external funding.

In securing sponsorship, academics may submit to highly prescriptive requirements on the objectives and forms of outcome of a project, as when receiving resources from a government department or private firm. Publication may be restricted. It is unlikely that the sponsors will seek to dictate the methods used, though that can happen in the social sciences when sensitivities or ethical issues arise in approaching and working with particular subject groups. Increasingly too research is driven by market considerations.

In some countries, but not all, researchers look for funds from private foundations whose demands on the objectives and forms of outcome of a project, once funded, are likely to be nonexistent or minimal. In the past, in the United Kingdom, the research councils were also regarded as a source of independent funding, although they varied: the former Agricultural Research Council acted virtually as the research arm for the Ministry of Agriculture. Researchers have increasingly moved from the responsive to the initiatory mode, and they are prescriptive about, for example, "researcher contact with user groups."

But where funding sponsorship has become more assertive on objectives and forms of outcome—where have methods, or epistemic characteristics, been affected? In the words of Elzinga (1985), has there been "epistemic drift"? Apparently not for the most part, however, (see our studies of the Foresight Initiative [Henkel, 2000; Henkel et al., 2000] on academic identities). It would be surprising if there had been, since sponsors cannot create knowledge that they themselves sponsor researchers to study the nature of knowledge—and create it for them.

Institutional Models

We can note several models of the relationships that convert forms of sponsorship into institutional formats:

1. *The autonomous individualistic model* which exists not only as, say, medical practitioners practising privately, or freelance journalists, but a minority of academics who have been able to escape the institutional co-op, perhaps by virtue of distinction, and to live perhaps on free grants but within institutional protection.

2. *The autonomous collegial model* is still the beau idéal and has as its premise that a group of practitioners will act to ensure their collective standards, by enforcing admission criteria, and will share certain resources, but will not exercise control, within broad limits, over the nature or volume of individual work. Its relationship with external sponsors is likely to be relaxed, though not necessarily at arms length, in that it is likely to rest upon established institutional protections from interference.

This model obtains in world-class universities, though it may now be increasingly mitigated by the numerical predominance of a second class of nontenured researchers and teachers and the increased dependence of even the most prestigious institutions on government or corporate funding.

3. *The managed model* is that which obtains in the private sector, and in some in-house units depending heavily on external sponsorship, where the objectives, methods, and format of outcomes are set managerially and directed toward ultimate application and profitability rather than to scientific ends, although deference to scientific codes of verification will be observed.

4. *The partnership model* where academics and industry reach agreements on quid pro quo.

These institutional ecologies may be both the products and the originators of particular power-knowledge mixes. The capacity to earn or the failure to earn different degrees of academic freedom will depend on various mixes of distinction and utility. The outcomes of the different forms are not easy to determine and differentiate. Power derived from teaching or research excellence may be enhanced by autonomy, but, equally, excellent research may derive from tightly managed centers. The power derived from perceived relevance is clearly demonstrable in some areas of technology, including clinical sciences and economics.

Nature of Resource Required

There is some literature on the effects of size of unit on both research functioning and economies of scale (e.g., Johnston, 1993; Kyvik, 1991, 1993). As far as institutional size is concerned, the jury is out on the economies of scale that are believed to level out as costs of coordination—particularly in multicampus sites—increase with size. It is assumed, at least by government agencies, and some megalomaniac heads of institutions, that quality follows size, though the reverse has often been true. (Consider the size of the University of Manchester between the wars which housed both Rutherford and Namier). In the United States, the "best" include both very large- and medium-size institutions. It has been generalized by Professor Peter Scott that "increased sizes lead to more bureaucracy."

In principle that is likely to be true, but if for this purpose we define bureaucracy as the dominance of managerial values and practices over academic managerial values and practices, we would need to compare, say, the Universities of Berkeley and UCLA with some tightly controlled former teacher colleges in both the United Kingdom and the United States. In general, therefore, size is an ambivalent characteristic that can affect academic power in different ways.

Stage of Finalization

It is tautologically evident that work, which has reached its final form, is more likely to secure both internal and external power than that which is still struggling to clarify its objectives, boundaries, and methods. At the intermediate stages, both objectives and methods may be more open to pressure or negotiation.

Examples of Knowledge-Policy Connections

Weiss (1977) has offered several clarifying relationships of these knowledge-policy connections. A general account of changing policy moods (Wirt, 1981) depicts a cyclical process in which public services might be set up and institutionalized so that power is exercised through dominant professions until the laity—politicians, interest and client groups—become dissatisfied and take power away from them. But, before long, replacement policies lead to new forms of professionalization and institutionalization, which perhaps a generation later will become challenged in their turn.

A U.K. example is that of the treatment of educationally impaired children. Under the 1944 Education Act, ten forms of "handicaps" were identified, and specialist schools and staff were created to attend to them. But with the Warnock Report (1982) and subsequent legislation, these categories were swept away in favor of generic "statementing" and instead a whole new profession of "special educational needs" was set up. With this have come new terminologies; new assumptions about the best ways of meeting needs; new legal stipulations and, of course, texts and training sequences. Their work has been named as "The only growth sector within education in the UK." A similar example might be the changing fortunes of public policies in the field of positive discrimination—the first phase being that of "color-blind" neglect, followed by a plethora of rules and legislation creating a race relations and antidiscrimination profession, and that to some extent followed by a reaction to these new forms of professional power, albeit reinforced by external reference groups.

The knowledge backing each of these policy phases will lie in the apparent ability to identify different forms of social or clinical impairment and create treatment structures for them. On these presumed abilities, both professional and legal determinism have been based. As the assumed knowledge base has changed so has the power derived thereof. Correspondence between different phases of policy development and knowledge styles have been noted.

Henkel (1998) has noted the ebb and flow of different conceptual and epistemological assumptions in public evaluation. There was an earlier shift from the positivist to the hermeneutic paradigm and "the associated change of emphasis to formative rather than summative evaluation." Within social evaluation, in the positivist phase, methods included the social survey, statistical analysis, and psychometric testing, and the preferred evaluative model was the randomized controlled experiment. But over time "awareness of the instability of social services undermined the claims of the experimental model. There was a shift towards description and the relationship between inputs, processes, context and outcome. Anthropological perspectives concerned with the interplay between milieu, process and inputs were advanced so pulling towards more context-specific approaches" (Henkel, 1998).

The objections to positivism with its search for regularities, systematic explanation, and prediction in social life have been well remarked: "People are not simply objects whose behaviour is in principle explicable in terms of a series of natural laws." "The concepts of intention, meaning and value are central to an understanding of human action and a grasp of them entails a comprehension of the language in which individuals and society express them.... The limitations of 'hard' scientific criteria become more strongly felt. Interpretative, illuminative, ecological and anthropological studies depending on internal logic rather than on external controls have intensified" (Henkel, 1998).

At the same time, we must be cautious about making global assumptions about these correspondences. For example, the styles attributed to positivist science—often used as a kind of liberal academic swear word—may be found in examples where knowledge has contributed to considerable human progress, including the reduction of privileged political or economic power. Medical epidemiological studies have been used to break, rather than advance, privileged hegemonies, as tobacco firms would ruefully agree. Whilst most educators would question the measurement and assessment of their performances against benchmarks and numerical scores of outcomes, some forms of connoisseurial inspection could be too subjective and biased and exercised in favor of particular educational doctrines. The tradition of Blue Book Exploration of social problems at the turn of the twentieth century was positivist in style but exercised the power of knowledge without any kind of institutional coercive framework. The knowledge was authoritative in that it could cause changes in behavior, but it did so by persuasion on key public issues, and in doing so it dislodged authoritative hegemonies.

The more recent history of both higher and lower school education in the United Kingdom shows well how different forms of evaluative knowledge seeking line up with assumptions about who should have the power and how it should be exercised. In the far past, higher education evaluation was not primarily hermeneutic in style but contained elements of both the summative and formative—depending on purpose and subject area, and administered by peer review that could be either exigent and external or connoisseurial and interactive. But the increasing desire of the state to break up academic hegemony, and to shift from standard setting by academics on their own criteria to standard setting on criteria set to public policy criteria, has led to drastic changes in the type of knowledge that is now created and employed. The state organizations assume that both teaching and research should have particular forms of outcomes that can be graded and thus measured, and that include contributions to the economy. The system is geared to ensuring that progress in achieving governmental targets can be measured and announced. The models of learning and research outcomes are tied to positivist assumptions about the efficacy of managerially endorsed criteria; academics and teachers are co-opted into the elaboration of the criteria that are, however, set as governmental a priori. The official knowledge is powerful because it is quantitative and, therefore, easily used for grading lists and summations and easily used to divert attention from the more subtle qualifications that apply to individual conditions and performance.

In the United Kingdom the return to positivism, which had begun to reverse to some extent from the early 1900s in school policy, has been decisive. It has become possible for the state to "know" the constituents of good education or research in schools and higher education, to know how to achieve them (through the pressures generated by outcome analysis, benchmarking, and associated rewards systems), and thus to know how to convert precise and quantified forms of knowledge into authoritative resource rewards and penalties. This assertion of arithmetical epistemics handily reinforces the shift toward managerialism at all levels of the system—managers can more easily use figures that are "thin" whilst words are "thick."

We may see in these examples a clear case of particular forms of knowledge seeking—public evaluation—responding to equally clear assumptions about the distribution and exercise of power.

Epistemics and Politics

Finally, we should consider the extent to which epistemological concerns and criteria are separable from political issues. It follows not only from the extension of academic boundaries explored by Trist (1972) with his domains subject to multiple reference groups, and Gibbons et al. (1994) Modes 1 and 2, but also from the fluctuating fortunes of positivism and interactive or hermeneutical versions of knowledge, as noted above, that these concerns and criteria are promoted partly out of the interior discourse of academics but also as part of largely political movements. The challenge of radical student groups to academic power in the 1960s and 1970s was part of a larger struggle for power, voiced largely as an attack on the authority of received knowledge, as indeed was academic resistance to it. Different forms of knowledge reinforce different philosophies of state and professional control as particularly exemplified in the remarkable story of the return of positivism in educational evaluation in the United Kingdom and elsewhere.

Yet few generalizations in this area are completely true or false. There remain academic groups who pursue internalist philosophies and practices in the certified surety that these remain the right way to advance knowledge. For the most part, they secure the best academic prizes and esteem that are cashable as grants, prestigious academic posts, and, in some subject areas, support and prestige in the outside world. At the same time, we have to note how some of the less rigorous academics have made their way into political influence by virtue of their communicativeness and perceived utility. Thus, we do right in trying to specify and generalize the "power-knowledge nexus," but remain tentative about any generalization derived from doing so.

Note

1. This paper was first delivered at a SCANCOR Conference on "Universities and the production of knowledge," held at Stanford University, April 25–26, 2003. In 2005 a revised version was published in *The Journal of Higher Education*, United States of America.

References and Works Consulted

Bardach, E. (1984) The Dissemination of Policy Research to Policy-Makers. *Knowledge* 6(2), December.
Bachrach, P., and M. Baratz (1970) *Power and Poverty*. New York: Oxford University Press.
Becher, T. (1989) *Academic Tribes and Territories*. Buckingham: Open University Press.
Biglan, A. (1973) Relationships Between Subject Matter Characteristics and the Structure and Output of University Departments. *Journal of Applied Psychology* 57(3), pp. 204–213.
Bourdieu, P. (1975) The Specificity of the Scientific Field and the Social Conditions of Progress. *Social Sciences Information* 14(6), pp. 19–47.
Buxton, M., M. Gonzalez-Block, M. Hanney, and M. Kogan (2002) *The Utilisation of Health Research in Policy Making: Concepts, Examples, and Methods of Assessment*. HERG Research Report, no. 28, October.
Caplan, N. (1977) The Use of Social Research Knowledge at the National Level. In C. H. Weiss *Using Social Research in Public Policymaking*. Lexington: DC Heath.
Cohen, D., and C. E. Lindblom (1979) *Usable Knowledge. Social Science and Social Problem Solving*. New Haven: Yale University Press.
Cronbach, L., and P. Suppes (eds.) (1969) *Research for Tomorrow's School. A Disciplined Enquiry for Education*. London: Macmillan.
Dahl, R. (1968) Power. In *The International Encyclopaedia of the Social Sciences*. New York: Free Press.
Elzinga, A. (1985) Research, Bureaucracy and the Drift of Epistemic Criteria. In B. Wittrock et al. (eds.) *The University Research System, the Public Policies of the Homes of Scientists*. Stockholm: Almqvist and Wicksell.
Fisher, H. A. L. (1935) *History of Europe*. London: Eyre and Spottiswood.
Foucault, M. (1977) *Power/Knowledge*. New York: Pantheon.
Gibbons, M. et al. (1994) *The New Production of Knowledge*. London: Sage Publications.
Henkel, M. (1991) *Government, Evaluation and Change*. London: Jessica Kingsley Publisher.

Henkel, M. (1998) Evaluation in Higher Education. Conceptual and Epistemological Foundations. *European Journal of Education* 33(3), September.

—— (2000) *Academic Identities and Policy Change in Higher Education.* London: Jessica Kingsley Publishers.

Henkel, M., S. Hanney, J. Vaux, Von Walden, and D. Laing (2000) *Academic Responses to the UK Foresight Initiative.* Research Report, Uxbridge, CEPPP, Brunel University.

Isaac, J. (2004) Conceptions of Power. In M. Hawkesworth and M. Kogan (eds.) *Routledge Encyclopaedia of Government and Politics.* 2nd ed. London: Routledge, pp. 57–69.

Johnston, R. (ed.) (1993) *The Effects of Resource Concentration on Research Performance.* Canberra: National Board of Employment, Education and Training.

Kogan, M., and M. Henkel (1983) *Government and Research.* London: Heinemann.

—— (2000) Getting Inside: Policy Reception of Research. In S. Schwarz and U. Teichler (eds.) *The Institutional Basis of Higher Education Research.* Chapter 2. Dordrecht Kluwer.

Kuhn, T. S. (1972) *The Structure of Scientific Revolutions.* 2nd ed. Chicago: University of Chicago Press.

Kyvik, S. (1991) *Productivity in Academia.* Scientific publishing at Norwegian universities. Oslo: Norwegian University Press.

—— (1993) Academic Staff and Scientific Production. *Higher Education Management* 5, July 2.

Lindblom, C. E. (1990) *Inquiry and Change. The Troubled Attempt to Understand and Shape Society.* New Haven: Yale University Press; New York and London: Russell Sage Foundation.

Lukes, S. (1974) *Power: A Radical View.* London: Macmillan.

Merton, R. K. (1957) *Social Theory and Social Structure.* Glencoe, IL.

Mulkay, M. J. (1977) Sociology of the Scientific Research Community. In Spiegel-I. Rosing and D. de Solla Price, *Science, Technology and Society.* London: Sage Publications.

—— (1979) *Science and the Sociology of Knowledge.* London: George Allen and Unwin.

Nagel, E. (1961) *The Structure of Science. Problems in the Logic of Scientific Explanation.* London and Henley: Routledge and Kegan Paul.

Nowotny, H., P. Scott, and M. Gibbons (2001) *Rethinking Science: Knowledge and the Public in an Age of Uncertainty.* Oxford: Polity Press.

OECD (1995) Educational Research and Development. Trends, Issues and challenges. Paris. Organisation for Economic Co-operation and Development.

Philp, M. (1985) Authority. In A. Kuper and J. Kuper (eds.) *The Social Science Encyclopaedia.* London: Routledge, pp. 55–56.

Polanyi, M. (1962) The Republic of Science: Its Political and Economic Theory. *Minerva* 1(1).

Popper, K. R. (1972) *The Logic of Scientific Discovery.* London: Hutchinson.

Rip, A. (1997) A Cognitive Approach to the Relevance of Science. *Social Science Information* 36(4), pp. 615–640.

Storer, N. W., and T. Parsons (1968) The Disciplines as a Differentiating Force. In E. B. Montgomery (ed.) *The Foundation of Access in Knowledge—A Symposium.* Syracuse: Division of Summer Sessions, Syracuse University.

Trist, E. (1972) Types of Output Mix of Research Organizations and their Complementarity. In A. B. Cherns et al. (eds.) *Social Science and Government. Policies and Problems.* London: Tavistock Publications.

Van den Daele, W., W. Krohn, and P. Weingart (1977) The Political Direction of Scientific Development. In E. Mendelsohn, P. Weingart, and R. D. Whitley (eds.) *The Social Production of Scientific Knowledge.* Vol. 1. Dordrecht: Holland/Boston USA: Reidel Publishing Company.

Warnock Report (1978) *Meeting "Special Educational Needs: The Report of the Committee of Enquiry into Education of Handicapped Children and Young People."* London, UK: HMSO.

Weingart, P. (1977) Science Policy and the Development of Science. In S. Blume (ed.) *Perspectives in the Sociology of Science*. Chichester: John Wiley and Sons.

Weiss, C. H. (1977) *Using Social Research in Public Policymaking*. Lexington: D. C. Heath, pp. 183–197.

Whitley, R. D. (1977) Changes in the Social and Intellectual Organisation and Social Organisation of the Sciences. In E. Mendelsohn, P. Weingart, and R. Whitley (eds.) *The Social Production of Scientific Knowledge*. Vol. 1. Dordrecht: D. Reidel Publishing Company.

———— (1984) *The Intellectual and Social Organization of the Sciences*. Oxford: Clarendon Press.

Wirt, F. (1981) Professionalism and Political Conflict: A Developmental Model. *Journal of Public Policy* 1, part 1.

Chapter Two
Universities and Society: Whose Terms of Engagement?

Mala Singh

Introduction

The idea of a socially engaged university belongs in a long line of moves to assign or appropriate the university for socially preferred purposes. Modernization, national development and nation building, manpower and human capital development, democratization and social transformation, and economic growth and competitiveness have been among the imperatives that have underpinned the arguments for the university to transcend its inwardly defined core functions of teaching, learning, and service and become more socially embedded. See Kerr (1995) on the Land Grant Movement of the 1860s in the United States; and war-related research at U.S. universities during World War II; or Coleman (1994) on Japanese universities in the 1880s supporting "modernization" through their teaching and research; and the political and human resource requirements in the Soviet model of universities. In the current conjuncture, the call for university engagement is part of the discourse of the "knowledge society," which has seen higher education assume a new prominence within the requirements of a "knowledge-driven economy" but also subject to a sharper accountability discourse driven by governments, global financial institutions, donors, and other social forces.

Universities are deemed to be critical to the knowledge society and economy but they are no longer upheld as the sole institutional location or agency for the production, use, and dissemination of knowledge (Gibbons et al., 1994). In this paradox of dethronement and restoration, what is the distinctive form of social engagement for the university in a knowledge society?

Thinking through this issue requires reconceptualizing not only the traditional missions, values, and functions of the university but also its familiar institutional forms and systemic locations and, most importantly, its relationships with an enlarged number of external constituencies. Universities in starkly different polities and economies are faced with the challenge of producing appropriate content for these reconceptualizations, as terminologies and discourses about engagement cross national, regional, and continental boundaries and become globally potent.

University engagement as a "black box" (Neave, 1998, p. 246) concept is easy to advocate, adopt, and celebrate. The view that "engagement with wider society" should be a core value for universities, that universities should respond to the needs and expectations of society and engage with multiple communities of interest has become quite commonplace in developed and developing countries alike and is, therefore, not seriously contested at the level of value or principle. It is the unpacking of the "black box" of engagement, both conceptually and empirically in the historical and geographical spaces that universities occupy that exposes the most stubborn and slippery normative and strategic challenges facing the translation of engagement into intelligible and sustainable contextual vocabularies.

In this chapter, I argue that, difficult as it is to act on and manage, the only notion of engagement that makes sense in the complex terrains of higher education in different parts of the world is a multidimensional one whose internal tensions and often unpredictable consequences require adroit steering and constant negotiation. Only a notion that tolerates and holds together a series of differently motivated interactions with external societal interests can do justice to the fact that higher education has multiple purposes and ends, not all reducible to narrow corporate understandings of the knowledge society. What is also to be considered in understanding the full measure of engagement are the interests of a range of internal and external stakeholder constituencies in the purposes, processes, and "products" of higher education and their different powers of leverage over them. I plan to interrogate the notion of engagement through juxtaposing higher education purposes and stakeholder interests in order to illuminate the terms of engagement, both in relation to the dominant social forces that are defining and driving engagement today and, more critically, in respect of other nonstate and noncorporate social interests that are often absent or only rhetorically present in the engagement debate.

I then go on to raise some questions about whether, and to what extent, a multidimensional view of social engagement is

- applicable beyond traditional public or nonprofit universities to the great variety of institutions, organizations, and arrangements that are now designated as higher education; and
- sustainable for universities in poor countries in the developing world, that are acutely constrained in their engagement choices by local socioeconomic and political impediments, and by the disadvantaged positioning of their countries and regions within the global "asymmetries of power,"

which make them more vulnerable to certain types of "reformist" discourses pertaining to economic liberalization.

I conclude by considering an overall normative frame that could be invoked—which goes beyond the purposes of higher education—to adjudicate contesting claims of different social forces when operating with a pluralistic notion of engagement of higher education. In grappling with the above issues, I seek to hold on to a Habermasian ideal in the midst of numerous struggles within and over the university by asking about the forms or models of engagement that would enable the university to deploy its considerable range of infrastructural and intellectual resources to benefit the society as a whole and yet carve out a space for critical debate and independent reflection on a host of social, political, and economic issues that shape and color our lives both locally and globally.

Engagement: Premises, Purposes, and Patrons

The engagement debate has to contend with the weariness (and skepticism) of many in higher education because of the ongoing intense scrutiny of higher education, both through external inspection and internal introspection. Universities have to negotiate their way across myriad, often contradictory, demands—to change radically in many respects and remain stable and consistent in others, to account to many stakeholders with hugely different needs and yet retain a recognizable measure of autonomy and independence, to become more individually competitive at national, regional, and international levels while operating in partnership and cooperative mode with other institutions, to compete successfully with vast commercial organizations that are becoming education vendors on an increasingly global scale, to increase access to formerly excluded student constituencies and improve the quality of provision with budgets that have not increased significantly, to retain a coherent identity and recognizable "brand" while decentralizing, outsourcing, and "unbundling," to be a space for critical and reflective thought while responding to the knowledge needs of industry and local communities, and to promote social justice and the public good in a global environment where the entrepreneurial pursuit of private goods is the norm. The path of engagement will have to be constructed through these many antinomies of demand for change and continuity.

Inayatullah and Gidley (2000, p. 1) argue that "the university stands at the gateway of a range of futures." How true is such a prognosis? For many, the fate of the university at the dawn of a new millennium appears to be already sealed, its future inextricably linked to and shaped by the "inevitabilities" of globalization. These inevitabilities include the usual suspects—the hegemony of the market and its package of values and priorities, the weakening of national sovereignties, the global dominance of organizations such as the International Monetary Fund (IMF) and the World Trade Organization (WTO), the powerful demands on knowledge organizations by the innovation needs of competitive economies, the "commodification of knowledge" (Schugurensky, 1999), and cultural homogenization through global media and communications technologies. For others, the future of the university is still unfolding, undoubtedly impacted on by the dynamics of globalization but open to carefully chosen interventions that seek to mediate some of the trajectories of globalization in order to better serve local purposes and needs. This approach to globalization and its impact on areas of social provision like education does not see globalization outcomes as inevitable (Mittelman and Othman, 2001, p. 7) or following a single preordained path or as a "unified" (Burbules and Torres, 2000, p. 13), symmetrical, uncontested, and context-free phenomenon. The unfolding of globalization is seen as holding spaces and opportunities for different outcomes, some of them potentially more emancipatory than others for larger numbers of people in developing countries who are currently excluded from many of the much heralded benefits of globalization.

If engagement is about negotiating a broad set of choices and directions for the university, it can only be within the context of a view that does not see the futures of the university as already fixed within a leveling globalization "teleology," notwithstanding the overt and covert ways in which globalization imperatives pressurize and homogenize higher education. Policies and strategies intended to give effect to the idea of an engaged university cannot but be located within the economic and political demands of globalization, especially in its neoliberal incarnation. But the eventual outcomes of engagement, whatever the intent, will depend on how the imperatives of globalization intersect with local conjunctures and how these are interpreted and layered onto existing institutional histories. The terms of engagement will be shaped by the strategic use of the opportunities and spaces afforded by the intersections of global and local pressures and will unavoidably include a variety of trade-offs made by the socially engaged university in order to succeed (or survive in some instances). This is an issue of more acute challenge and consequence for crisis-ridden universities in many poor countries.

The initiative of the Association of Commonwealth Universities (ACU) to provoke "worldwide debate among some five hundred member universities" (Coldsteam and Bjarnason, 2003) on the issue of university engagement is a powerful reaffirmation of the broad social purposes of higher education in a context where narrowly economic purposes are imposing their dominance. It is a call to universities to take on the accountability imperative proactively and preemptively, and move the debate about the future of higher education beyond defenses of failing models or critiques of developments in higher education that offer no feasible alternatives. In an argument that fits well into the ethos of self-regulation, it proposes engagement motivated not by a form of Kantian hypothetical imperative but a categorical imperative that is premised on what is rational and right for universities to do and legislated largely by and for universities themselves. In a project that aspires to a global reach, the arguments about the scope of engagement in the Consultation Document, especially the checklist of "indicators" for "assessing progress" in institutional engagement, have the potential to shape and give direction to conceptualizations and practices of engagement in a number of contexts, countries, and regions worldwide, even where there are no Commonwealth universities.

The opportunity and danger in this initiative to "universalize" certain core understandings of engagement lies in the fundamental premises on the basis of which one mobilizes around engagement. It is obviously preferable to aim at the institutionalization of the most conceptually and contextually nuanced versions of engagement and at weakening its worst reductionism forms. In the search for the "fundamentals" of engagement, the balancing act of institutions in accommodating the *pulls* and *pushes* of stakeholders with varying powers of "persuasion" should be seen as an enduring existential challenge rather than as a settled issue that privileges some possibilities of engagement and closes off others. Conceptualizing engagement as an ongoing struggle to accommodate contending normative and strategic demands may help to focus universities on the varying purposes of higher education, and the rationales and drivers of engagement associated with those purposes. It may also make it possible for the Association of Commonwealth Universities (ACU) Framework for Engagement to assume the higher moral ground in conceptualizing and steering the "future" of the university in a more socially nuanced way when viewed against other global developments impacting on higher education. These include, for example, the World Bank's New Framework for Higher Education that will directly impact on all countries applying for bank loans for tertiary education reform, and the World Trade Organization's (WTO) Framework for the General Agreement on Trade in Services that seeks to bring higher education services *under the auspices* of international trade agreements. These last two frameworks also seek to bring higher education into a larger social arena but their underlying values, political premises, and economic conditionality—linked as they unambiguously are to market efficiencies and freedoms—are threatening to a fuller notion of human and social development.

The approach to engagement of the ACU's Consultation Document is appropriately comprehensive in scope, both in relation to the range of societal stakeholders as well as to the types, levels, and objectives of interactions. The formulations open up an ambitious and generous space for the practices of university engagement, one that is potentially enabling and beneficial for the multiple purposes of higher education and for different actors and stakeholders within and outside the university. "Every institution has already developed working connections with policy-makers, industry and commerce, local communities and the wider society; none starts off with a clean slate. The whole web of these interactions and the setting of university policy to foster them are what we term *engagement*" (ACU, 2001).

Engagement is a comprehensive term for all aspects of a university's policy and practice. It "implies strenuous, thoughtful, argumentative interaction with the non-university world in at least four spheres: setting universities' aims, purposes and priorities, relating teaching and learning to the wider world; the back-and-forth dialogue between researchers and practitioners; and taking on wider responsibilities as neighbours and citizens" (ACU, 2001). The Consultation Document makes it clear that basic and applied knowledge, high-level skills for social and economic development, and constant responsiveness to societal needs and requests for new programs and services are now required of universities, irrespective of country context or circumstance. What is also important to note is that engagement is not seen merely as one noble initiative among any number of others. Neither is it a random collection of ad hoc and disparate activities. The ACU position on engagement is a comprehensive and maximalist one, argued for as the very raison d'être of the university.

With this in mind, engagement covers a huge canvas of university activities and operations and has many interlocutors and addressees. But not being an end in itself, what academic and societal ends can, and should, engagement serve? A possible answer lies in the way one disaggregates engagement. This, itself, can be done in different ways! One way is through looking at its meanings and implications for university goal setting in relation to the traditional core functions of teaching, research, and service. One can consider engagement according to university location and function in the global or regional or national political economy. One can look at its implications for different institutional types in higher education (public, private, face-to-face, distance, electronic, etc.) or the different ideological roles and identities associated with universities as a "site of dissent . . . a corporation (run as a business), . . . as a place for academic leadership (teaching and additive knowledge), . . . as providing the ideological legitimacy of the state, . . . as a public service (the university that exists for the community)" (Inayatullah and Gidley, 2000, p. 226). One can link engagement to the aims and purposes of higher education, or to the interpretations and expectations of key actors, stakeholders, and beneficiaries within, and external to, higher education. These and, no doubt, other possible routes of disaggregating could all produce useful understandings to illuminate engagement in its many complexities and contradictions.

I want to look at engagement vis-à-vis the multiple purposes of higher education and associated stakeholder interests in these purposes. The

"Education White Paper 3: Transformation of Higher Education in South Africa" (1997) is a good example of a national policy framework for higher education restructuring that lays out a number of different but related purposes for higher education, in this case linked to the needs of social reconstruction and a better quality of life for all in a posttransition society.

The White Paper expresses the following, in order:

- To meet the learning needs and aspirations of individuals through the development of their intellectual abilities and aptitudes throughout their lives, higher education equips individuals to make the best use of their talents and of the opportunities offered by society for self-fulfilment. It is thus a key allocator of life chances, an important vehicle for achieving equity in the distribution of opportunity and achievement among South African citizens.
- To address the development needs of society and provide the labour market, in a knowledge-driven and knowledge-dependent society, with the ever-changing high-level competencies and expertise necessary for the growth and prosperity of a modern economy. Higher education teaches and trains people to fulfil specialized social functions, enter the learned professions, or pursue vocations in administration, trade, industry, science and technology and the arts.
- To contribute to the socialization of enlightened, responsible and constructively critical citizens. Higher education encourages the development of a reflective capacity and a willingness to review and renew prevailing ideas, policies and practices based on a commitment to the common good.
- To contribute to the creation, sharing and evaluation of knowledge. Higher education engages in the pursuit of academic scholarship and intellectual inquiry in all fields of human understanding, through research, learning and teaching. (The Education White Paper 3, 1997, pp. 7–8)

Economic growth and prosperity and labor market needs are identified as necessary purposes for higher education but so are the facilitation of equity and the development of an enlightened and responsible citizenry. Individual aspirations for intellectual development and upward social mobility are flagged but so are the requirements of the "common good." The White Paper seeks to embrace the multiplicity and diversity of ends and objectives in higher education, covering the spectrum of what is both valuable and necessary for a higher education system to do and to aspire to, both educationally and socially. As necessary a spectrum as this list covers—given South Africa's task of social reconstruction—it immediately exposes the huge challenges in meeting the objectives of the whole ambitious package in a single system or worse still, in single institutions, particularly in a context of limited resources and capacity, and competitive behaviors for market share.

Many higher education policy frameworks as well as institutional mission statements contain similar mixes of purposes, embracing societal needs that range from the hard-nosed economic to the more intangibly social and civic.

It is the latter that mostly tend to lose out in the engagement debate through being treated as "good to have" ends that are ambitious sometimes even achievable (perhaps serendipitously) but that are not quite in the same plane as pragmatic "necessary to have" outcomes. The latter often carry associated incentives and disincentives that concentrate the engaging intent better than the former could. Benefits to industry or "the economy" are increasingly expected as direct outcomes to be delivered by universities while benefits to other noncorporate stakeholders continue to be viewed as indirect possibilities, trickling eventually down unspecified societal pathways to address the more abstract purposes of higher education.

Making the notion of engagement more explicitly practical requires greater attention to strategies, "indicators" and proxies for evaluating activities and interactions in relation to the more abstract purposes of higher education. If not, the content of engagement will remain largely shaped by what is easier to evaluate and quantify, driven by powerful stakeholders (including governments and the corporate sector) who strongly prioritize knowledge and skills for economic growth and competitiveness rather than by civil society constituencies whose priorities may require knowledge and skills that could facilitate greater levels of democratic consolidation, social cohesion, social justice, and other such values and objectives encapsulated in the broad purposes of higher education.

On the one hand the range of purposes in higher education point to different spheres of societal existence that are sometimes in alignment but, on the other hand, very often in contention with one another. The nature of the knowledge society desired by innovation-hungry multinational corporations or tolerated by insecure governments may not include the kinds of knowledge needed by critical citizens who may want to engage with the negative impacts of the global dispensation of power and privilege or with local and regional policy choices and behaviors. However, even as a jostling unruly package of tasks, the purposes of higher education point to different but equally important dimensions of social need and aspiration and the measure of the high expectations of higher education in addressing these needs and fulfilling these aspirations.

The contending nature of the multiple purposes of higher education is nowhere more evident than when one seeks to relate the purposes of higher education to the range of stakeholders interested in how a university conceptualizes and acts on engagement, stakeholders with sometimes overlapping, sometimes quite sharply exclusive interests and needs. The ACU Document points to many of the key stakeholders: government, industry and commerce, health and education leaders, administrators, planners and student representatives, the professions, employers, local communities, taxpayers, and voters. Responding satisfactorily to the expectations of this range of stakeholders (who even within their own categories will not reflect homogenous positions on their expectations) is a tall order for universities as for any other social institution.

Two points need to be made in this regard. First, the engagement debate cannot be silent about internal role players such as academics and researchers. To view them primarily as the "delivery agents" of the engaged university removes the fundamental stake that they have in the engaged university. They are not stakeholders in the way in which industry or local communities are, but they are critical to how sophisticatedly engagement is interpreted and how willingly it is implemented. They certainly have a stake in the nature and consequences (both intended and unintended) of the reconciliation of academic values with external social demands, in how the multiple purposes of higher education are to be held together in teaching and research, and in how to reconcile conflicting priorities and maintain academic integrity. In the paradigm of a university as a communicative and interactive space for and with multiple stakeholders (Delanty, 2001) one has to ensure that their role is not confined to being only the technically expert interpreters of the needs of other stakeholders. Their own particular stakeholder identity in the process of defining and negotiating the parameters of engagement must be factored into the stakeholder equation. This does not exclude the possibility that many of them will comfortably ally their academic and personal interests with those of powerful paying external stakeholders. On the other hand, some of their interests will surely also include a broad defense of academic ideals or emancipatory social goals within the fray of engagement. Whatever the interests of such internal actors might be, they should not be underplayed or excluded in the continuous negotiations over the nature and terms of engagement.

The second point in regard to contending stakeholder interests has to do with the, involvement and role of external nonstate and nonmarket stakeholders. What leverage, capacity, and knowledge do they have to direct the university's attention and resources to their needs, and to shape the university's response accordingly? Guy Neave reminds us that the word "stakeholder" encompasses actors with different powers in relation to what is at stake. "As Napoleon the Pig observed in Orwell's satire 'Animal Farm,' so with stakeholders and partners—some are indeed more equal than others" (Neave, 1998, p. 247). Barnett also points to the more compelling leveraging power of corporate stakeholders over others in shaping the content of engagement: "Engagement can take many forms, but some will be promoted more vigorously than others. The big battalions will favour the more performative versions of engagement: the university will be persuaded to structure itself in favour of activities likely to have exchange value in the knowledge economy" (Barnett, 2003, p. 138). In the battle to shape the terms of engagement, there is hardly likely to be a "level playing field" in stakeholder power and influence.

In a context of acute pressure on universities to develop additional sources of income, to cultivate industry partnerships, and to find sustainable markets for their knowledge "products," universities can be easily persuaded that their primary engagement interlocutors (and beneficiaries) should be from the corporate world. There are undeniably benefits from that interaction for other stakeholders as well. Higher education and industry collaboration could internally lead to curriculum innovation and renewal, and to new research directions and sources of funding. Externally, it could provide possibilities for local or regional development or increased employment opportunities in niche areas. But what are the possibilities that civil society stakeholders or small local communities will be considered as critical partners in university engagement, especially where they have little or no purchasing power over knowledge products or expertise, or have knowledge needs that, if addressed, may generate tension for the university with corporate engagement partners? The principle of cross-subsidization offers an obvious way of supporting activities relating to societal stakeholders who are unable to be paying "customers" but only if the university recognizes its responsibility to take account of the needs of noncorporate stakeholders in its circle of engagement. The problem becomes more complicated when other interested parties want access to the knowledge "products" commissioned from the university by corporate stakeholders. What is the responsibility of university researchers and administrators if an environmental impact study commissioned by a petroleum company reveals massive threats of environmental degradation and dangers to the health of poor communities living in the vicinity? How does a university juggle its engagement responsibility to industry partners and to community development in such a situation where it is aware of the possibility that dominant market interests are likely to "trump up" those of all others if left unmediated?

To discharge its responsibility as a "discursive community" and provide a communicative space for stakeholder interaction, the university could facilitate a dialog between different stakeholders whose claims to knowledge and its ownership conflict with each other. In fact, the provision of this kind of discursive forum has been identified by Delanty as the distinctive feature of the university in a knowledge society. Hitching its star instrumentally neither to the state nor to the market nor to internal academic interests, the university functions in a way that "mediates, or interconnects, several discourses in society" (Delanty, 2001, p. vii). For Delanty, a defining part of the university's transformative mission and role lies in its functioning as a Habermasian public sphere in order to "expand reflexively the discursive capacity of society and by doing so to enhance citizenship in the knowledge society" (Delanty, 2001, p. vii). In operating in this way, the university enhances both the democratization of knowledge as well as the participatory capacity of citizens to deal with global cultural and technological forces. From the point of view of differentially empowered stakeholders in the engagement debate, a "communicative understanding of the university" presumes that the university has a responsibility to "become the clearing house for all the voices that would otherwise be silent or muted beyond recognition"

(Delanty citing Fuller, 2001, p. 156). To the many interpretations of university engagement with and for citizens and communities must be added this dimension of enabling civil society discourses and their associated interests to assert their claims and concerns in a context where other more powerfully driven discourses are seeking to shape the meanings of the knowledge society.

Functioning as a discursive space for stakeholder claims will not always spare university participants from entering the debate as affected and interested parties since the dispute is also about conflicting claims on the university itself. This is especially true in a situation where the corporate partner is a reluctant participant and not forthcoming enough in terms of its own social responsibilities. The university will be confronted with having to reconcile norms and strategies that have been driven apart by engagement dilemmas, and make tough political, economic, and moral choices in order to ensure that the values contained in the purposes of higher education, the ideals contained in the university mission, and social goals pertaining to democracy and justice are not compromised fatally in the cut and thrust of engagement. This could occasionally require the university to be more interventionist in translating its broader social responsibility into choices that contribute to the weakening of the asymmetrical power relations among different stakeholders in the engagement arena, thus mediating their negative consequences as much as is possible.

I would like to make one last point about "citizens" as stakeholders in and partners of the engaged university. The Association of Commonwealth Universities (ACU) Consultation Document highlights the important role of citizens in shaping the nature and scope of university engagement, seeking to extend the traditional way of thinking about "community service" as one of the core functions of universities. Community service usually includes the provision of continuing and adult education opportunities, making available university resources and expertise for the support and development of local communities, and in some instances "service learning" arrangements for credit-bearing student participation in projects agreed to with communities. The Consultation Document argues for a view of engagement that moves citizen involvement much more centrally into the operations of the university, right from the point of setting university purposes and priorities to exposing research results to "public debate" beyond peer and expert review. The centrality of citizen involvement in higher education clearly makes for a more socially aware and socially connected university. It also brings in train all the complex problems of the appropriate structural and process arrangements for such involvement in university planning, stemming from the differential powers, limits, and responsibilities of the role players in the interactions of university constituencies and citizens (Muller and Subotzky, 2001, pp. 163–182).

In all of this, there is the problem that "citizen" may too easily be thought of in terms that do not go beyond the immediacy of local communities.

Substantial university involvement in local community development needs would already be an enormous gain in the trajectory of university engagement but this does not exhaust the possibilities of engagement with citizens and communities. The parameters of engagement must include wider conceptualizations of social groups—viewed locally, nationally, regionally, and globally—and take on board issues that have to do with, for example, the requirements of distributive justice, the causes and consequences of asymmetrical power relationships between strong and weak economies, and new forms of marginalization and impoverishment fostered by globalization. All of these impact on the meaning and possibilities of citizenship and community development at very local levels and need to be addressed in order to give full effect to the ACU's view that the essence of engagement is "taking the world with full seriousness" (ACU, 2001, p. 38).

Although it is clear that context and circumstance will shape the particular histories of engagement, I would like to end this section by taking stock of certain critical *starting points* and *core premises* that could enable engagement to fulfill its potential to transform universities on the basis of a greater social connectivity that is intended to maximize the conditions for human development in a knowledge society:

The engagement debate is not foreclosed by the power and impact of globalization on higher education. Some of the space to manœuvre may be narrowed down, even more so in the margins of the global political economy, but, as has been pointed by many globalization theorists, other spaces open up for transformatory action. Certain types of normative and strategic choices can and should be exercised in order to maximize progressive outcomes from different forms of engagement. The realization of any particular set of possibilities depends on the purposes and terms of the engagement between higher education and society, on the power of the actors who shape or influence these terms, and on the interplay between institutional policies and politics on the one hand and national or regional political and economic circumstances on the other.

The parameters and possibilities of engagement can be more concretely understood when related to the purposes and ends of higher education. These purposes are manifold, embracing different kinds of societal needs ranging from economic growth and human resource development to strengthening democracy and the values of social justice. The strategies for engagement must be genuinely open to as many of these pathways as are urgent and necessary in particular social contexts.

To be most useful and least distorted, engagement should be understood as pluralistic and multidimensional. Tension may exist among the different dimensions, and they may not often be reconcilable in a smooth and complacent view of engagement. The way of being for the university is a constant struggle to balance different forms of engagement rather than a settled view of engagement driven by or effectively serving only the most powerful stakeholders. A complex view of engagement can also function as a yardstick by which to criticize and correct one-sided or reductionist views of engagement.

The engaged university has to ensure that the full range of stakeholders is included in its societal interactions, not only those who are the most economically powerful and clamorous. For universities the interactions with corporate stakeholders as a definitive component of engagement is clearly necessary and unavoidable in the current conjuncture and carries as many exciting possibilities for university transformation as worrying ones. But the participation of noncorporate stakeholders has to be an equally definitive dimension of engagement in order to bring in the full range of societal interests on which to engage. A Rawlsian argument about whether and how engagement serves the interests of the worst-off in society may be useful to invoke in establishing the range of stakeholders and beneficiaries of university engagement.

In reaching out to citizens, engagement cannot be reduced to community development in ways that do not bring into the picture the global issues pertaining to the asymmetrical distribution of power and privilege. The cosmopolitan and internationalist role of higher education requires that the gaze of engagement is global as much as local, combining moral and intellectual concerns about the nature of the global society that is emerging and its differential impact on local communities in the various countries where Commonwealth universities are located.

Engagement for Universities or for all Higher Education Providers?

My argument about engagement is, first, that its scope and reach is better understood when seen against the requirements of the multiple purposes of higher education and their associated societal goods, and second, that one of its primary challenges lies in accommodating the full range of stakeholder interests in the purposes of higher education. Can such a view of engagement be applied to the variety of institutions and arrangements commonly designated as higher education, or does it make most sense in relation to universities alone? Many of these new forms of higher education do not include what are often understood to be defining elements of higher education, for example, a spread of qualifications across a range of faculties and subjects, some relationship between teaching and research, the provision of senior postgraduate qualifications especially at doctoral level, some activities of community service, and so on. The question about the applicability of engagement begs a prior one as to whether the different purposes of higher education can be achieved within single institutions, even universities, given the current climate of demand and constraint facing most of them. This problem is particularly acute for institutions that may be the only universities in small poor countries. Mario et al. point to the tough challenge facing Eduardo Mondlane University in Mozambique in trying to reconcile its "role as the country's only full-blown university with demands for immediate market relevance" having to compete with single-purpose for-profit institutions while also driven by "ideals of service and community development" (Mario et al., Partnership for Higher Education in Africa, 2001, p. 63).

Under the pressures of massification, the skills needs of the knowledge economy, developments in information and communications technologies, and the thrust toward the view of education as a market, one of the most striking developments in higher education has been the proliferation of organizational forms and modalities for higher education provision. Public and private universities and colleges, corporate universities for in-house workforce education and training, for-profit and nonprofit institutions, face-to-face, distance, electronic and mixed-mode provision, universities that combine research, teaching, and service, institutions that do no or little research, and those that offer only career-oriented programs in one or a few fields (e.g., business management and information technology) all lay claim to being higher education institutions. This development is hailed as proof of the innovative and exciting directions in which higher education is moving beyond the missions, identities, and modalities of the traditional university. Sometimes, within the celebratory accounts of new developments in higher education, there are brief, often unelaborated, cautions about the need for values in education that are required for living in democratic societies amid the seductions of technological connectivity, "virtuality" and competition-driven market responsiveness in higher education (Salmi, 2000, p. 3). These cautions are a good opening to reflect on the broader societal responsibilities of the new institutional types in higher education.

A hundred flowers are already blooming in terms of institutional typology but there is also the dizzying phenomenon of increasing virtuality in higher education, understood not only in terms of campus-eliminating electronic provision but also the trend toward the "unbundling" of a number of higher education activities and functions, usually held together within a single institution. The scope of such unbundling covers not only the separation of the often associated functions of teaching and research but even functions within teaching and administration, for example, program design and delivery, assessment, and certification, or "registration, payment and student recordkeeping" offered as separate packages by different service providers (Newman, 2000, p. 3). Efficiency and cost become the prime criteria for evaluating a range of unbundled functions. In such a context of "unbundling," it is not clear who, if anyone, will be answerable for the issue of societal engagement. The issue of engagement may very well falter on the scattering of functions implied in the trend toward unbundling as well as toward "coalitions, partnerships and networks." Scott points out that, as traditionally constituted institutions and systems decline, the problem may "not come from new kinds of 'university' but from no kinds of 'university'" (Scott, 1999, p. 9).

The issue of engagement responsibility also raises its head in relation to higher education institutions offering qualifications across national borders. Do transnational providers have any responsibilities for interacting with citizens in local community development issues in the different countries and regions where they now operate? Recent debates and initiatives (UNESCO, 2002) about the possibilities of international regulation of transnational providers of higher education have focused on the issue of codes of conduct for such providers. Their concerns have primarily and rightly been addressed to whether the quality of provision of what is offered abroad is equivalent to the quality of provision in the country of origin of the provider, whether the curriculum seeks to be reasonably context-sensitive and whether the appropriate formalities of local regulation have been observed. The requirements of societal engagement on transnational providers within the growing phenomenon of "borderless" education remains an issue to be addressed in broader discussions about globalization and higher education. Its resolution is likely to be related to the debate about whether, in the light of the demands of a *global knowledge society*, it makes sense to think of higher education as a *global public good*. If the answer is affirmative, it would require all forms of higher education, not only universities, to be accountable for promoting such a global public good within the context of their own particular missions, rather than operating within the dichotomy to which Van Damme (2002) draws attention that public universities serve public and private good purposes but that for-profit private higher education institutions serve only private interests.

One approach to the engagement conundrum lies in following the argument of those who seek to distinguish sharply between the university and other forms of higher education provision, many of which offer career-oriented programs responsive largely to the training needs of the private sector. Altbach has argued that such for-profit institutions or arrangements should not be designated as universities or offer academic degrees since they do not engage in research or community service, offer specialized competency training only in certain market-related fields, and indicate no interest in, or responsibility for, public good issues (Altbach, 2001, p. 2). On this definitional reordering of higher education institutions, the obligations of engagement could be made to apply only to universities. If one wanted to render this positively, one could argue that the vocational and commercial missions of the other provider types, especially for-profit providers, automatically contribute to economic (therefore social) development and require no further indicators of societal engagement. On a negative reading, one could suppose that those provider types, especially in their lower level training role, lack the tradition, the gravitas, the capacity, or the inclination to contribute to broader social goals through research or academic expertise or public good commitments. Attractive as this view might be for those disturbed by the appropriation of the university "brand" (Scott, 1999, p. 8), it presumes too tidy a world where traditional universities will remain on their side of the definitional fence or that nontraditional providers will cede the ground that they have already occupied in the university domain.

Exemption from the social engagement imperative for more privatized forms of higher education removes any pressure on them to give considered attention to the forms and parameters of the societal obligations that relate to their educational functions or even their commercial missions. Perhaps the least unsatisfactory and most open-ended starting point in such a complex and evolving issue is to think of different, perhaps lesser and greater, forms of engagement for different types of higher education providers rather than a total exemption for any single type. Institutions constituted as universities have a clear obligation to foster engagement on a number of societal fronts, stemming from their wide missions and their multifaceted activities, many of which are supported by public funds. But so do other institutions, including for-profit ones, which are located within communities of need and which prepare large numbers and diverse categories of students for passage into the economy, society, and community. The development of criteria and indicators for these different institutional forms of engagement and some form of monitoring could become an interesting debate within the jurisdiction of quality assurance and accreditation, especially if one wants to ensure that teaching and learning, research and community service are all reconstructed through the prism of engagement. This would give quality assurance a more explicit role in university transformation but only on the proviso that quality assurance became much more socially reflective and less bureaucratically self-referential.

Engagement: Contexts and Conjunctures

To what extent is it possible to develop a common conception of the fundamentals of engagement that will have the same resonance for the large number of universities operating in different parts of the world, in countries with different forms of insertion into regional and global economic and political blocs? Given the relativizing impact of history, geography, and economic conjuncture and without much more research on actual contextual realities, it may be really difficult to make nontrivial observations about university engagement that make sense across contexts and continents without significant qualifications of all kinds. Most higher education institutions are having to engage with the demands of radical change and increased responsiveness in their own societies, operating in contexts where local demands are made more acute by the pressures of a globalizing economy. What, for example, could be the content of engagement for universities in sub-Saharan Africa in countries whose populations, viewed on a global scale, are the most impoverished and endangered in relation to the basic necessities of human survival and dignity? The enormous scale of human struggle with poverty, disease, drought, famine, civil war, political authoritarianism, and decades of debilitating structural adjustment programs provides obvious ground for social engagement by universities that represent resources of infrastructure, knowledge, information, expertise, agency, and activism, no matter how meager or impoverished they themselves might be. But what are the real possibilities to develop and sustain appropriate policies and practices for engagement that do not become captive to exclusively entrepreneurial rationales and drivers in situations of extreme funding and resource constraint?

The current debate about engagement, especially as it might apply to African universities, has had powerful predecessors. Debates about universities and their social responsibilities had already been waged in many African countries in the postindependence period of the 1960s. Political leaders, communities, donors, intellectuals, academics, and students were all enthused by the idea of the "developmental university" as part of the development armory of the new nation-state. The terms of engagement were quite explicit, Nyerere's well-known formulation is blunt about the fact that universities were instruments for national development: "The University in a developing society must put the emphasis of its work on subjects of immediate moment to the nation in which it exists, and it must be committed to the people of that nation and their humanistic goals. We in poor societies can only justify expenditure on a University—of any type—if it promotes real development of our people. . . . The role of a University in a developing nation is to contribute; to give ideas, manpower, and service for the furtherance of human equality, human dignity and human development" (cited in Coleman, 1994, p. 335). Apart from the political unit of reference (the nation-state), and despite the difference in context and period, the ACU Document expresses similar sentiments about what society expects of universities: "The world depends increasingly on universities for knowledge, prosperity, health and policy-thinking. Universities are thus required to become engines of development for people, institutions and democracy in general" (ACU, 2001, p. i). In order to develop an appropriate understanding of university engagement within the requirements of the knowledge society, it is necessary to interrogate the continuities and discontinuities between the current "developmentalist" notion of the university as it might apply in sub-Saharan Africa and the conceptualizations and practices of the earlier postindependence period.

Previous debates in other developing country contexts about the societal responsibilities of universities have contained similar sentiments to the African expectations of universities. A review published in the mid-1970s of selected higher education institutions and their impact on national development in Asia and Latin America and Africa makes for familiar reading. "In the developing countries, the focus of development is determined by the rural character of the societies, their relative underdevelopment, and the necessity of grappling with such rudimentary needs as food and nutrition, public health, low per capita incomes, unemployment or underemployment, weak spots in the education system, the preservation of cultural values, and the movement towards equity and social or ethnic equality. The real issue is what higher education can do and what it is doing about these fundamental needs" (Thompson et al., 1976–1977, p. 6). Also emphasized is the necessity for higher education to support nation building, build national leadership, and foster cultural cohesion as well as cultural tolerance.

In an analysis that is still pertinent, the Latin American Report sets out the parameters for engagement in identifying five "major capacities" for social improvement against which the contribution of higher education could be judged: "(1) the capacity to comprehend, define, and give priorities to society's needs and aspirations; (2) the capacity to comprehend and define the problems that arise in the process of meeting these social needs and aspirations; (3) the capacity to formulate various alternatives for solving problems; (4) the capacity to apply appropriate technology to alternative solutions; (5) the capacity to select and apply the various mechanisms, strategies, and policies to solve these problems" (Thompson et al., 1976–1977, p. 191). In the regional reports from all three continents, the approach is similar: higher education "in the service of development can help increase the capacity of nations to produce more... material goods... but it must also reinforce concern for justice, morality, and human dignity" (Thompson et al., 1976–1977, p. 167). The scope of impact-related expectation includes an increased measure of economic goods but equally strongly emphasizes the importance of social, political, and ethical goods in the social engagement of universities. These broader social goods are no less critical today for university engagement. The challenges of poverty and unemployment, of social exclusion, of access to human rights and many other necessities highlighted in the earlier period of development discourse continue and persist but in a way that is now complicated by the framing power of new global discourses of accountability whose particular understandings of knowledge and competence may be too limiting for a comprehensive view of human and social development as envisaged in both earlier and current debates.

From the experience of the past of both positive and negative impacts of requiring universities to become instruments and agents of national development, there are many cautionary lessons for the contemporary debates on engagement as the raison d'être of universities. Taking into account some of the key lessons of history with regard to social engagement, how can universities in sub-Saharan Africa make sense of the current imperatives of accountability and societal responsiveness in a context that is paradoxically similar in respect of wretched living conditions and meager opportunities for vast numbers of the population, and yet different in relation to the discourse of the knowledge economy in a globalizing world, the weakened role of the nation-state, and the power of the market in shaping social development? "What possibilities of normative and strategic choice for engagement are there for universities that have themselves been devastated by funding neglect due to sharp reductions in state and donor support, 'exploding enrolments' which put a strain on physical infrastructure like classrooms and residences, teaching capacity and quality of provision, overcrowded libraries with dated holdings and general deterioration of working conditions and demoralisation of staff all round?" (Sawyerr, 2002, pp. 23–24).

Developing and implementing a pluralistic view of engagement requires some fundamental enabling conditionalities. Critical among these are a state that has some capacity and commitment to provide and regulate in relation to higher education. "Even where state funding support levels are drastically reduced and where private sector and user fees form increasing proportions of new sources of university revenue, the state's responsibility for critical infrastructural and core costs at least, is abdicated at great peril" (Sawyerr, 2002, p. 59). The absence, paucity, and unreliability of public funds make the university more vulnerable to market- and donor-driven imperatives and less able to set and follow a coherent and multidimensional engagement agenda for effective education, research productivity, and societal outreach. Some element of sustainable state funding (which is, itself, not too crudely tied to rate-of-return expectations) could allow for agenda setting and interpreting engagement in ways that are responsive to the huge social, political, and economic needs of poor communities and societies. Conditionalities for university reform linked to funding provided by donors and international lending organizations also need to be flexible enough to allow for university choices that give effect to a pluralistic view of engagement.

A regulatory role for the state is also critical in order to implement an engagement agenda that is wide-ranging and beneficial to different stakeholders and participants in higher education. Even where state-funding support levels are low, the responsibility for some oversight in relation to issues of access and equity for students, and the quality of education and training should vest primarily in the state, especially in contexts where universities have embarked on various forms of internal semiprivatization of university functions in order to generate badly needed operational income, or where there is a rise in the number of private providers. The absence of such oversight allows for the possibility that equity of access is undermined by a user-fees approach and that the quality of provision is compromised by rapid enrollment growth, poor staff capacity, and outdated library and other facilities, and that transformation goals in education and in society are made more difficult by the prioritization of market values. The state could play an important role, for example, in ensuring that transnational for-profit providers of higher education operate within acceptable quality and access requirements and have some measure of articulation with local state-supported institutions.

The presence of institutional leadership driven by a complex view of engagement that takes account of local, regional, and global factors is also critical for any discussion of the engaged university. Visionary and skilled leadership is needed, ideally at different levels of the university's operations to balance the entrepreneurial side of the university with social and intellectual imperatives, and to balance the academic priorities of the university with the huge developmental demands made on the university. In a context where state support or capacity is minimal or absent, the value of wise and capable institutional leadership becomes even greater.

For many universities, the above enabling confluence of external and internal factors is not likely to materialize, creating a hydra-headed challenge for those who have already embarked on university reform in countries that have undergone processes of political and economic liberalization since the 1990s. The attempts at university regeneration and reform in countries like Ghana, Kenya, Mozambique, Tanzania, and Uganda, have been captured up in a series of case studies (Manuh et al., Partnership for Higher Education in Africa, 2003) whose fuller analysis could yield more information on the explicit and implicit forms of societal engagement embedded in the revitalization measures undertaken at those universities. The authors provide accounts of remarkable initiatives to resuscitate and reform a number of universities, many of which had flourished with generous state and popular support in the pre-1970s period and then undergone severe crisis and destabilization through funding slumps, increased enrollments, political interference, and academic depletion. Almost all the authors identify ongoing pressure on university finances, lack of planning and management capacity, academic staff in need of retooling, threats to quality from rapid student enrollments, the deterioration of the learning and living environments in relation to libraries, laboratories, and residences, poor development of information and communications technologies, the decline in the research ethos, the inability of public universities to compete with private providers, and the threats to equity and social justice as a result of admissions and appointments driven by market considerations, as some of the recurring challenges facing African universities as they embark on internal reform.

For many of these universities, the most radical first step in such a disabling context might very well be a set of initiatives for the building of basic institutional capacity and stability, in this way establishing a platform to conduct and facilitate more effectively the traditional functions of teaching, learning, and research. The negative impact on academic quality of years of financial constraint, unregulated expansion, and hasty and unprepared program responsiveness has been identified as a key problem in many of the case studies. Addressing this challenge would already be a huge step forward in relation to the primary responsibility of the university to produce knowledgeable, employable, and socially aware graduates. This will in any case bring in train internal struggles over policy control and direction as well as resources. However, the processes of curriculum reform, of seeking new student constituencies, of developing a new research agenda for the university, and of identifying new sources of funding would all require the university to go down an engagement trajectory that could enrich and enlarge the revitalization process that is underway in many countries.

The need for more reliable research data and rigorous interrogations of reform and revitalization initiatives at universities in many sub-Saharan African countries has already been identified in recent analyses of African higher education (Sall et al., 2002; Sawyerr, 2002). Such research and analyses will clearly provide a finer-grained sense of contextual specificities that is necessary for fleshing out the nature, limits, and possibilities of engagement for universities and other higher education institutions on the continent. In anticipation of more contextualized understandings of the possible trajectories of engagement, some *preliminary observations* can be made in relation to the engagement debate:

1. What will be the impact of the knowledge society discourse on engagement initiatives in universities in sub-Saharan countries? The worrying implications for higher education of the global frameworks and arrangements developed by the World Bank (WB) and the World Trade Organization (WTO) assume greater resonance for universities in poor countries that will be unable to mediate the conditionalities of globally lending organizations or international trade agreements as they impact on national policy. The influence of such frameworks in directing the content of engagement along narrow one-dimensional conceptualizations of the knowledge society could be disastrous. An unreconstructed concept of the knowledge society could become a poisoned chalice for universities seeking to understand the comprehensive requirements of societal engagement in what constitutes, arguably, the most marginalized and impoverished sections of the global population. Any discussion on the parameters of an emancipated form of engagement for universities in poor countries will, therefore, have to be premised on an interrogation of what the knowledge society means in real terms for those societies and whether the notion of the knowledge society in its current incarnations is sufficiently enabling for key dimensions of human development to be advanced. Academics and intellectuals, as well as institutional and national policy makers, whose options and choices will be directly impacted on by the forms and content of particular conceptualizations of the knowledge society, should play a key role in setting the terms of this interrogation.

2. Since one is not starting the university engagement debate in a vacuum or with a clean slate, how can one usefully link the current debate about engagement with earlier conceptions and experiences? On what can one build? Moreover, what should provide cautionary lessons for the future? Sall et al. point out that even in the decades of crisis and degradation universities in sub-Saharan Africa were relevant to their societies in a number of different ways. Public universities were "key sites for debate, critique and mobilization on behalf of political change" (2002, p. 2). They were also instrumental in satisfying a continuing social demand for higher education as enrollments snowballed in the 1980s and 1990s in spite of the grim counterfactual reality of diminishing employability, especially in the public service. Both these dimensions hold huge possibilities for university engagement in the current situation. Manuh et al. (2003) point to the necessity for universities to "chronicle and analyse" reform initiatives in Ghana and their implications for and impact on individual and institutional life, including the effect of structural adjustment programs and economic liberalization on universities themselves. Providing independent evaluations of the efficacy of poverty reduction strategies chosen by government is seen as a critical task that could be undertaken by universities in a more systematic and institutionalized fashion. This in turn could enable university constituencies like researchers, unions, and students to "be proactive in using research and information on public policy measures to support advocacy that could raise the profile of universities as beacons of knowledge, linked to action" (Manuh et al., Partnership for Higher Education in Africa, 2003, p. 114). Clearly, issues of political atmosphere, policy research capacity, and institutional willingness to link university reform more explicitly with social and political reform will all play a role in the extent to which universities will or can play the kind of socially engaged role advocated above.

3. In relation to the issue of social demand, universities that seek to locate their reform and revitalization initiatives within the frame of social engagement will not be able to avoid the challenges of demand, access, and equity. Almost all the case studies have pointed to the trend toward market-driven admissions in higher education and the access problems posed by this trend for particular categories of students (women, poor students from outside the main cities, etc). Given the low levels of participation in higher education, and limited opportunities for higher education study in the face of enormous demand, universities will exacerbate "imbalances of socio-economic class, gender and regional origin in the student population" (Mario et al., Partnership for Higher Education in Africa, 2001, p. 61) unless they themselves are able to insert mediating mechanisms to open up access, for example, "affirmative action" for underrepresented student constituencies (including the disabled), loan and scholarship programs (Musisi and Muwanga, 2001, pp. 36–37), "external degree centres and distance learning initiatives" to reach students in more remote parts of the country (Manuh et al., Partnership for Higher Education in Africa, 2003, p. 122).

Conclusion

I have argued for a "messily" multidimensional notion of university engagement. It is a notion that invites a variety of rationales and content and requires visionary management of contradictions, benefits, and dangers through a cunning mix of dialog incentives, and stipulations. It requires frequent translation from an easy conceptual plane into the thornier realm of sensible policies and flexible strategies in order to reshape many aspects of institutional culture. If managed well, governance and leadership styles, decision-making structures and processes, the equity and diversity profile of staff and students, staff expertise, and curriculum and research innovation may all see transformative changes that help to move the university forward in its own development. But what are the larger politics that inform the requirements of university engagement?

Looking at engagement through the prism of the different purposes of higher education and their associated stakeholder interests made it possible to introduce one layer of norms by which to judge how engagement should be conceptualized, managed, and held together. It may be necessary to go beyond this layer—which speaks largely of what happens within higher education—to a larger normative frame that pertains to the role of different kinds of social institutions, including universities, in advancing human development. Without such an external frame, the university could interpret the pluralism of engagement as a series of efficiency-driven compromises to ensure that it does not lose is financial benefits whenever conflicts of interests have to be resolved. A postmodern understanding of pluralism will not sufficiently enable the university to consider that some societal interests are more rationally and morally compelling than others and some engagement choices more emancipatory than others. Within the discourse of the knowledge economy, the growing power of corporate interests in the work of the university will create many dilemmas for universities that still want to maintain some "public good" commitments in and through their work.

What happens in the engagement arena when the interests, claims, and pressures of different societal groups cannot be managed or reconciled, and also cannot continue to be endlessly juxtaposed in a communicative roundabout? (Barnett, 2003). Such situations may require the university to take a position for the sake of its own institutional integrity in holding on to a larger conception of societal values and ideals that affect the lives of a large number of people. By what yardstick can a judgment be made either by the university itself or by other interested parties that it is discharging its responsibilities as a major social institution committed to enhancing the many substantive requirements of a democratic and caring society?

I want to invoke Sen's analysis of the relationship between institutions and freedom in order to address some of the metapolitical issues surrounding the question of university engagement. The yardstick that emerges from this analysis is the extent to which institutions enable and advance human freedoms. "Individuals live and operate in a world of institutions. Our opportunities and prospects depend on what institutions exist and how they function. Not only do institutions contribute to our freedoms, their roles can be sensibly evaluated in the light of their contribution to our freedom. To see development as freedom provides a perspective in which institutional assessment can systematically occur" (Sen, 1994, p. 142). The question for the engagement debate concerns the larger contribution of the engaged university to "development" understood in Sen's sense as "a process of expanding the real freedoms that people enjoy" (Sen, 1994, p. 3). Such an expansion is influenced by "mutually reinforcing connections" between "economic opportunities, political liberties, social powers, and the enabling conditions of good health, basic education, and the encouragement and cultivation of initiatives" (Sen, 1994, p. 5). The yardstick for university engagement emerging from this conception may not be as philosophically abstract as it looks. It is, in fact, already embedded in the engagement responsibilities of the university toward some of the addressees identified in the ACU Document, for example, *students* and *citizens*. On this perspective, it should be possible to ask, for instance, whether the university is making it possible for students to acquire the full range of knowledge, skills, and understandings required by them to enable them to maximize their freedoms not only as job seekers but also as individual human personalities, as confident citizens of their countries, as empowered members of their communities, and as informed "global citizens" entering debates beyond their national borders? How have the infrastructural resources of the university, its research findings, its expertise structure impacted on the quality of life of citizens and communities, judged not only against an "aggregated" benchmark of economic growth but also against concrete "social indicators" relating to progress of individuals and communities against the debilitating conditions of "poverty, unemployment, inequality" (Coleman, 1994, p. 341 citing Seers). How has the university sought to connect the needs of students and citizens/communities in order to maximize benefits to both and enable them to enjoy different kinds of freedom relating to social and political empowerment as much as providing opportunities for economic development?

The normative frame of expectation from universities provided by the above perspective makes it possible to both enlarge and sharpen the "indicators" for judging engagement progress. It also provides an external reference point by which to orient the policies, strategies, and practices of engagement, especially where powerful stakeholders can tip the scale when irreconcilable conflicts of interests occur. Such a normative frame clearly contains assumptions about the responsibilities of major social institutions to enhance democratic participation as much as to foster economic development, and about the obligations of the university to the realization of the "public good," especially in relation to equity and social justice.

But is there a real role beyond rhetoric for such a normative frame to shape the direction and content of engagement in the busy operational life of the university? The strategies and practices of engagement may only tangentially relate to the values and ends contained in the normative commitments of a higher education institution. As university leaders and administrators know well, the strategic life of the university is vulnerable to the logics of powerful and demanding external stakeholders whose normative frames of reference may differ in key respects from those of key role players of the university or other less powerful external stakeholders. In reality, the university is likely to be driven by conjunctural pressures and demands such as competition for students, declining state and donor funds and the search for "third stream" contract income through research, consultancy, and other services, the pressure to account to government authorities for the use of public funds, the pressure to maintain reputational competitive edge, and an endless round of such drivers.

The requirements of the knowledge society are ostensibly what underlie the renewed attention to the university as a socially engaged institution. But if engagement is to be not only comprehensive but transformative as well, both in its internal institutional reconfigurations and in its external societal impact, it must encompass and advance values and goals that relate to the many dimensions of human development. For this to happen, the terms of the knowledge society will themselves have to be emancipated from the monopolistic demands of the market, and reconceptualized to include political, social, and ethical considerations that are currently absent or only weakly gestured to.

Within such an enlarged view of the knowledge society, the continuous alignment of the strategic life of the university within normative frames of reference that cover different kinds of social goods can become more institutionalized since the "indicators" of the engaged university in a knowledge society will require it. Otherwise, the language and practices of engagement will privilege some dominant interests over others in ways that are bound to impoverish the engagement project as a whole.

References and Works Consulted

ACU (Association of Commonwealth Universities) (2001) *Engagement as a Core Value for the University: A Consultation Document*. London: ACU.

Altbach, P. (2001) The Rise of the Pseudo Universities. *International Higher Education* 25, pp. 2–3.

Barnett, R. (2003) *Beyond all Reason: Living with Ideology in the University*. Buckingham: SRHE and Open University Press.

Burbules, N., and C. A. Torres (eds.) (2000) *Globalization and Education: Critical Perspectives*. New York: Routledge.

Coldsteam, P., and S. Bjarnason (eds.) (2003) *The Idea of Engagement: Universities in Society*. London: Policy Research Unit of the ACU.

Coleman, J. S. (1994) The Idea of the Developmental University. In R. L. Sklar (ed.) *Nationalism and Development in Africa: Selected Essays, University of California Press*. Reprinted from Atle Hetland (ed.) *Universities and National Development: A Report of the*

Nordic Association for the Study of Education in Developing Countries. Stockholm: Almqvist and Wiksell International, pp. 85–104.

Delanty, G. (2001) *Challenging Knowledge: The University in the Knowledge Society.* Buckingham: SRHE and Open University Press.

Education White Paper 3 (1997) A Programme for the Transformation of Higher Education. Department of Education, South Africa.

Fuller, S. (1999) The Governance of Science. Buckingham: Open University Press, p. 98.

Gibbons, M., C. Limoges, H. Nowotny, S. Schwartzman, P. Scott, and M. Trow (1984) *The New Production of Knowledge.* London: Sage.

Inayatullah, S., and J. Gidley (eds.) (2000) *The University in Transformation: Global Perspectives on the Futures of the University.* Connecticut: Bergin and Garvey.

Kerr, C. (1995) *The Uses of the University.* Cambridge, MA: Harvard University Press.

Manuh, T., S. Gariba, and J. Budu (2003) Ghana's Publicly-Funded Universities. Case study funded by the 4-Foundation Partnership.

Mario, M., P. Fry, L. Levey, and A. Chilundo (2001) Higher Education in Mozambique. Case study funded by the Foundation Partnership.

Mittelman, J.H., and N. Othman (eds.) (2001) *Capturing Globalization.* New York: Routledge.

Muller, J., and G. Subotzky (2001) What Knowledge Is Needed in the New Millennium? *Organization* 8(2), pp. 163–182.

Musisi, N. B. and N. K. Muwanga (eds.) (2003): Makere University in Transition 1993–2000. Published in association with Partnership for Higher Education in Africa, James Currey, Oxford and Foundation Publishers: Kampala, p. 61.

Neave, G. (1998) On the Naming of Names. *Higher Education Policy* 11(4), pp. 245–247.

Newman, F. (2000) Saving Higher Education's Soul. *Change*, Washington.

Sall, E., Y. Lebeau, and R. Kassimir (2002) The Public Role of the University in Africa. Background document for the SSRC/AAU Project on the public role of African universities.

Salmi, J. (2000) Facing the Challenges of the 21st Century. *International Higher Education* 19 (Spring), pp. 2–3.

Sawyerr, A. (2002) *Challenges Facing African Universities.* Accra, Ghana: AAU (Association of African Universities).

Schugurensky, D. (1999) Higher Education Restructuring in the Era of Globalization. In R.F. Arnove and C.A. Torres (eds.) *Comparative Education: The Dialectic of the Global and the Local.* Oxford: Rowman and Little Publishers.

Scott, P. (1999) Globalization and the University. Keynote speech at CRE 52nd Bi-Annual Conference, Valencia, October 28–29.

Sen, A. (1994) *Freedom as Development.* Oxford: Oxford University Press.

Thompson, K. W., B. R. Vogel, and H. E. Danner (eds.) (1976–1977) *Higher Education and Social Change: Promising Experiments in Developing Countries.* 2 vols. New York: Praeger.

UNESCO (2002) *Global Forum on "International Quality Assurance, Accreditation and the Recognition of Qualifications."* Paris: UNESCO.

Van Damme, D. (2002) Outlook for the International Higher Education Community in Constructing the Global Knowledge Society. Presented at the UNESCO Global Forum on "International Quality Assurance, Accreditation and the Recognition of Qualifications." Paris: UNESCO.

Chapter Three
Knowledge, Globalization, and Hegemony: Production of Knowledge in the Twenty-First Century

Paul Tiyambe Zeleza

Introduction

As we begin the new century, indeed the new millennium, it is tempting to crystal-gaze into the future and reflect on the past; to see ruptures of old ends and new beginnings, with anxiety or anticipation. We hear about the emergence of new knowledge economies and new economies of knowledge, although it is not always clear what these terms mean—to which knowledge and to which economy they are most applicable. The changes that are taking place in knowledge and production systems, both real and rhetorical, are marked by spatial, social, and institutional dynamics, in that they manifest themselves unevenly between countries and regions, social classes and groups, and the institutions of knowledge production themselves. This is to caution against narratives, condemnatory or celebratory, that homogenize developments (economic, political, social, cultural, or ideological) taking place in our exceedingly complex and unequally integrated world—as is so common with the universalizing but exclusionary discourse of globalization.

The aim of this paper is to offer broad reflections on the production of knowledge in the twenty-first century—specifically the intersections between knowledge, globalization, and hegemony. At one level, the task may appear relatively straightforward, for these terms have become part of the contemporary social and political vocabulary. Yet, they are complicated concepts over which much intellectual sweat has been spilled and careers have been made and ruined. In a presentation such as this, it is only possible to give the barest of explanatory notes on what I understand by each of these terms, and the implications, in their interaction, for understanding the daunting challenges that knowledge production systems, especially universities, currently face—and are likely to face—as the new century progresses.

The paper seeks to discuss trends in various parts of the world—with a particular emphasis on Africa, which I have studied extensively. Appropriate comparisons are drawn in several places with trends in the countries of the global North, especially the United States, the country where I have lived and worked for many years and with whose higher education system I am quite familiar. Such comparisons are useful because they enable us to transcend the common myopic tendency of enveloping developing and developed countries in the mystifications of exceptionalisms. Comparative analysis not only underscores interconnections between universities in the global South and the global North, but also the fact that these universities are confronted with broadly similar challenges—even if from different vantage points, and with varied institutional capacities to manage them.

The presentation is divided into four parts. First, I will outline the debates about globalization, knowledge society, and hegemony. Second, I will explore some of the implications of globalization on higher education, historically the most important site of knowledge production, which is now facing enormous pressure. Third, I will examine the impact of the changes, currently taking place in universities, on knowledge production and academic freedom understood broadly to mean the capacities of academic institutions, groups, and individuals to produce critical social knowledge. Finally, I will briefly look at the manifestations of these trends, specifically in Africa.

Some Notes on Globalization, Knowledge Society, and Hegemony

Intellectuals and policy makers, like ordinary people, have their fads by which they frame their aspirations and anxieties, express their fantasies and fears. Globalization is the discursive craze of our times—although it may already be losing its luster, even among its staunchest proponents from the fallout of the September 11, 2001 terrorist attacks in the United States. Needless to say, there are many views about globalization, about its efficacy as a concept and effects as a set of conditions. Depending on how it is defined and perceived, globalization has its advocates, adversaries, and ambivalents. According to Petras (1999), the advocates and beneficiaries of globalization are found among the ascending countries and technocrats, the dominant economic enterprises and commercial classes, while the adversaries are concentrated in the dominated countries among peasants, workers, and small businesses. Those ambivalent about globalization consist of classes and enterprises that both win and lose from specific policies. Each tendency has its intellectual protagonists.

To its advocates—the hyperglobalists—globalization is seen as a new phenomenon involving a fundamental restructuring of the global system that is both inevitable and irreversible; a new global socioeconomic system is supposedly emerging from the crumbling old order of accumulation, social organization, and state sovereignty. To the antagonists—the skeptics—there is really nothing new about globalization, for it looks and smells like the age-old world capitalist system, with its insatiable capacity for conquest, domination, exploitation, and the production of inequality, disorder, and crises; it is, they aver, contingent and susceptible to interruptions and ruptures as has happened to previous globalization cycles. To the ambivalents—the transformationalists—the contemporary wave of globalization surpasses that of earlier epochs in terms of the extensity of global networks, the intensity and impact of global interconnectedness, and the velocity of global flows; it represents a historically unique confluence in various domains of social, economic, cultural, and political life.

It seems to me that we need to differentiate between the historical and ideological registers of globalization and the highly uneven and unequal links of various regions with it. Thus, the real controversy is an ideological and analytical one on how to periodize globalization's origins and trajectories, characterize its technological, economic, cultural, and political dynamics, and assess its impact on different classes, communities, countries, and continents. Understood as a historical process, referring to international or transnational communications and exchanges of capital, commodities and cultures, ideas, images, iconographies and institutions, practices, peoples, plants and places, and values, visions, vices, and viruses—then the world has been globalizing for a long time, although the process accelerated rapidly during the course of the twentieth century. Whether seen as a process or a project, a description of present conditions or a prescription for particular futures, the globalization industry has privileged perspectives and positioning from the global North, with the global South peeping in from the sidelines or the streets that wreck the meetings of the international financial institutions and the Davos global elite.[1]

Opposition to globalization as it is currently configured or conceptualized is certainly strong among African policy makers and intellectuals. Amin (1997, p. 2) regards globalization as the manifestation of contemporary capitalism's chaos, which "is visible in all regions of the world and in all facets of the political, social and ideological crisis." He insists that globalization via the market is a reactionary utopia that must be countered "by developing an alternative humanistic project of globalization consistent with a socialist perspective" (Amin, 1997, p. 5). Nabudere (2000, p. 11) has argued that contemporary globalization is neither new nor "a unilinear phenomenon proceeding in some predictable fashion, progressively moving from one state to another on a kind of evolutionary grid. On the contrary, it is recognized that globalization now proceeds unevenly in the same way that capitalist modernization proceeded in the earlier phases, even within Europe." For Africa, globalization has meant structural adjustment programs, which have derailed postindependence development efforts and have led to what he calls the "third colonial occupation," distinguished by the downsizing of the postcolonial state and downgrading of democracy. The disastrous consequences of contemporary capitalist globalization for Africa are outlined in graphic details by Mkandawire (1997a, 1998; Mkandawire and Soludo, 1999, 2003), who vigorously contests the neoclassical explanations of the African economic crisis of the 1980s and 1990s, which seek to absolve external factors, principally the international financial institutions, and blame the crisis on internal factors, such as poor natural conditions and human resources and political conflicts, and especially policy failures.

For many African intellectuals, therefore, globalization is seen as a destructive phenomenon and a coercive ideology from the global North that, despite some of its novelties, is only slightly different from previous forms and phases of capitalist imperialism. In the words of Ake (1995, pp. 22–23), "[I]t is not an abstract universal that is magically emerging everywhere; it is concrete particulars that are being globalized. What is globalized is not Yoruba but English, not Turkish pop culture but American, not Senegalese technology but Japanese and German." For Ake, then, "uneven globalization is not only a process but also an ongoing structuration of power . . . [it] is the hierarchization of the world—economically, politically, and culturally—and the crystallizing of a domination. It is a domination constituted essentially by economic power" (Ake, 1995, p. 23). Aina (1997) states unequivocally that African intellectuals should beware of Northern discourses of globalization.

"Just as globalization is about power relations and the construction of hegemonic order, [Northern] analyses rely on constructs that reflect and express a view and realization of that power, of that world, of those who construct it, and the place from which they perceive it. . . .[T]he globalization theories imagine and envision the world within a limited scope which is place-determined in terms of privileging a particular Eurocentric (Northern) positioning or understanding which undervalues, ignores or rejects non-European, non-Northern visions and knowledge. Backed by the very global power being studied," she notes wryly, "these discourses succeed in imposing on the rest of the world, particularly the South, their outline of the visions and imaginations of the globe. This in itself reflects the uneven power relations we are talking about" (Aina, 1997, p. 19).[2]

Similarly problematic are the concepts of knowledge and knowledge society. Although the concept of knowledge society has gained currency in recent years in academic, public, and policy circles, the idea of knowledge society can hardly be new, since knowledge has always been central to human existence and has played a role in all phases of historical development in every society. Academic definitions of the knowledge society are often framed in epistemological, sociological, and economic terms. The epistemological debates center on the meanings, forms, and claims of knowledge, in which various binaries battle it out for supremacy (scientific and ordinary knowledge, scholarly and social knowledge, explicit and implicit knowledge, codified and tacit knowledge, experiential and reflective knowledge, theoretical and practical knowledge, and constructivist and objectivist views of knowledge).

Sociological and economic literatures tend to focus on the changing relations between science and technology, knowledge and industry, knowledge and information, the increasing knowledge intensity of economic activity, the emergence of knowledge as the fourth factor of production, the growth of knowledge-based companies, the rise of postindustrial societies dominated by a new class of knowledge workers and professionals, "an expertoisie" as someone calls it—an ugly word that should be banished before it takes root—that has eclipsed the old divisions between the bourgeoisie and workers of industrial capitalism, the development of postmodern consciousness with its reflexivities and multiplicities of identities and multitasking needs, and the proliferation of knowledge production sites away from the universities. Much of this is indeed happening, but it is not happening with the same levels of intensity or directionality everywhere, and a lot of it is not new.[3]

Hegemony has also become the flavor of our times. An idea initially rescued by the Left from Antonio Gramsci in the 1970s to liberate Marxism from its arid economic determinisms, it has become popular to all and sundry and now refers to all manner of power relationships involving hierarchies and domination. As with globalization, there is little agreement on its primary referents (the state or class), on its analytical spheres (political, economic, or cultural), and on the levels at which it is exercised (national, regional, and global). There is heated debate on the interplay between coercion and consent in how hegemony is exercised, on the legitimacy and illegitimacy of hegemonies, and the nature, dynamics, and directionality of hegemonic cycles, transitions, and projects as well as counterhegemonic projects and contestations.[4]

Those who stress the cultural dimensions of hegemony tend to focus on the ideological and discursive practices of power, the permeability between dominant and subordinate cultures, and the ambiguity of consent. They often differentiate between the physical and mental forms of power, physical coercion and symbolic consent, social meaning and material reality, the different realms of behavior and consciousness in the everyday exercise and effects of power on political subjects who, even in their subalternity, seem to retain a capacity to resist domination and to reproduce domination in their resistance.

For others hegemony is more fruitfully analyzed in terms of the interstate system, but they do not agree on what leads to the rise and fall of hegemonic states, the identities of the hegemonic powers at different moments in world history, and continuities and discontinuities in the structure of hegemonies. The world system theorists believe that while hegemonic and counterhegemonic states have come and gone, the world capitalist system has remained hegemonic globally for at least the past 500 years, if not longer, while others argue that there has never been a systemic hegemony in the entire world system, although not for lack of candidates; that hegemonic stability has never existed at the level of the entire world system; that indeed hegemony is rare and transitory, it is never complete, and it generates opposition that undermines it.

Let me summarize these brief definitional notes. Hegemony is located in the complex, multilayered, hierarchical, and interactive structures of society, economy, politics, and culture, at various geopolitical levels, mediated by force and ideology, as well as the social inscriptions of class, gender, race, and religion. Capitalist globalization as a process and an ideology of restructuring the social relations of production and global hierarchies constitutes the hegemonic force of our age, against which many counterhegemonic struggles, forces, and movements are poised.

By globalization, then, I understand it to mean the contemporary processes of global capitalist restructuring, which are underpinned by neoliberal ideologies and policy interventions that are known in the global South, certainly in Africa, by the infamous name of structural adjustment programs (SAPs). My argument is that the regime of structural adjustment is global, although as with previous cycles of global restructuring, it is the weaker countries and classes that are paying the price with their sweat, security, and even lives. For these countries the old humanistic and historical dreams of anticolonial nationalism for development, democracy, and self-determination, or to put it differently, for developmental democratic states and sustainable development are still compelling, rather than the current rhetoric of knowledge economies and societies dominant in the global North.

Capitalist globalization entails the liberalization and privatization of the economy and public goods, including education. Thus, it is an ideology that seeks to impose neoliberal discipline on institutions of higher education, which affects all aspects of the academic enterprise, including teaching, research, and service. This, inevitably, has a profound impact on knowledge production and academic freedom. My focus on knowledge, then, is confined to scholarly or academic knowledge, or knowledges produced in research settings, which include and have historically been dominated by colleges and universities. The question, then, is: How is the hegemonic discourse of globalization affecting systems of knowledge production?

Globalization and Higher Education

As I have already indicated, stripped of all the controversies, the much-trumpeted claim that we live in the age of globalization simply refers to the intensification of international connections, contacts, and communication, and the growth of a more interdependent world integrated by new information and communication technologies. However, it is a process fraught with contradictions. The globalization of financial markets and multinational corporations is accompanied by economic regionalization and informalization; cultural transnationalism spreads simultaneously with rising cultural chauvenisms and fragmentations; states everywhere (and especially in the global South) are said to be declining as they proliferate and increase their internal repressive capacities; the certainties about the materiality of globalization are trailed by crises of knowledge about the world in which we live; and global terror, perpetrated by secretive states and shadowy organizations has done much to temper the sense of delight or dismay, apparent only a few years ago that the juggernaut of globalization was unstoppable save for some divine intervention. Above all, while globalization has transformed many old spatial, social, and institutional hierarchies and hegemonies, it has reinforced some of them and constituted new ones.

Universities have a peculiar relationship with globalization for as institutions they see themselves as universal communities of ideas; they trade or aspire to trade in international intellectual currency. It would not be an exaggeration to argue that research conducted by the world's universities has helped produce globalization as a constellation of material and imaginary, spatial and symbolic processes, while globalization is simultaneously producing new contexts and imperatives for intellectual communities. In other words, universities are impacted by globalization and are implicated in the discursive framing of globalization in that they have always aspired to be globalized, and they indeed themselves are globalizing institutions. Yet, there is a widespread sense that universities as institutions and academics as a profession are facing unprecedented crises engendered by globalization; that rapid technological, economic, political, and sociocultural transformations emanating from the wider world and academe itself are eroding the old systems, structures, and stabilities of higher education. Powerful internal and external forces that are as much pedagogical and paradigmatic as they are pecuniary, political, and demographic are reconfiguring all aspects of university life constituted around the triple mission of teaching, research, and service. Struggles of various kinds and intensities are being waged within and outside university systems over their missions and mandates, legitimacy and status, as leading producers, disseminators, and consumers of scientific and scholarly knowledge.

Part of the analytical challenge is that it is not always easy to disentangle developments caused by the contemporary processes of globalization, however defined, and those that may have other causes, or merely reflect age-old tendencies for institutional, intellectual, and ideological transformation. The literature is full of paradoxes: colleges and universities are experiencing rapid growth in student numbers in the face of declining public funding, as is the case among public universities in most American states and most African countries[5]; they are seen as central to the needs of the knowledge economy, but in many countries including (some in the OECD) they receive ever-declining public research funding[6]; they are prey to too much and too little state intervention; greater emphasis is placed on equity and access, while costs explode; there are tensions between traditional modes of teaching and career expectations of students, between flexibility of learning and standardization of courses, between providing critical knowledge and imparting information and credentials, between old notions of scholarly production and new forms of academic performativity, between the proliferation of publishing outlets and the tighter hold of prestigious journals as a screening mechanism for jobs, promotion, resources, and reputations; universities are research institutions producing an ever-declining share of research compared to business enterprises, as other players (including public and international agencies) expand their research activities; more people are involved in knowledge production but more of it is privatized; the academic profession has never been larger and appeared more endangered and casualized; pressures are intensifying for restructuring universities into unified national systems and differentiating between and within them; the expansion of universities and the scientification of society are accompanied by the decline in the socioeconomic status of universities and ambivalences toward science; and universities still like to see themselves as belonging to an international community of scholars while they are being turned into an export industry.[7]

A lot indeed is happening to institutions of higher learning as centers of knowledge production. These changes are of course connected to contemporary transformations both in society and the academy, for universities, even the lofty ones, cannot escape the push and pull of their societies. Their operations and practices are also often inscribed by an intricate interplay of institutional, intellectual, ideological, and individual locations and predilections. In order to get a better handle on the changes that institutions of higher learning are undergoing as centers of knowledge production, I have identified seven key trends dubbing them, if a little colorfully, the seven Cs: corporatization of management, collectivization of access, commercialization of learning, commodification of knowledge, computerization of education, connectivity of institutions, and corrosion of academic freedom. These trends are not new of course, but they have become more urgent and more complex, they are combined in new ways, and they manifest themselves unevenly in different world regions.

Corporatization of management refers to the adoption of business models for the organization and administration of higher education institutions. Universities are being pressed into the discourse of accountability and entrepreneurship, obliging them to undertake new budgetary strategies and to expand and diversify their sources of funds to become more efficient, productive, and relevant. Critics, mostly to be found among leaders of student and staff unions or associations interested in promoting broad representation and accountability by university management and all those wedded to the old ideals of a liberal education, point out that the reigning ideology of free-market capitalism increasingly sees education, not primarily as a social good or as a human right, but as an economic investment, and universities are turned into mills to produce and retool entrepreneurs and information operatives, instead of oases to nurture the values of democratic citizenship.

By collectivization of access, I mean the growing massification of higher education, the perception that education is a lifelong learning process, and the increasing collaborations among universities, and/or interventions in university affairs by stakeholders in the public and private sectors, which has resulted in the reconfiguration, or erosion, of traditional notions and values of university autonomy, academic freedom, liberal education and quality. The upsurge in higher education reflects the expansion of the youth population, the growth of middle-class incomes and aspirations, the creeping credentialism in professions and occupations, and the rising demand for knowledge-based skills and jobs. Given the rapid economic changes, the separation between education and career as chronologically distinct phases of life are crumbling. Thus, universities are adapting to the demands of continuing education for workers in the knowledge-based industries by restructuring their courses, making them part-time and modular. Consequently, universities are becoming more diversified in their programs and student composition. Already, in some countries, older working students outnumber younger students.

Commercialization of learning refers to the rapid expansion of private universities, the increased involvement of private enterprise in the provision of higher education, and the establishment of "executive" programs in public universities. Thus, we are seeing the rise of what some call the "market-oriented university," the "entrepreneurial university," or the "consumer university." Besides the spectacular growth of private universities related to the rising demand for higher education, and the ever-changing needs of the knowledge-based economy, corporate universities are emerging. These universities are created by large industries or transnational corporations.

These changes are eroding the monopoly that universities have long enjoyed over the resources and privileges of knowledge production as it has spread to numerous private and public sectors, including business, government agencies, and civil society organizations seeking the social legitimation conferred by recognizable competence. The connections among these institutions are exceedingly complex, but partnerships are being formed, indeed encouraged, and many academics rotate among them with varying degrees of comfort and ease.

All these developments reinforce the commodification of knowledge, as reflected in the increasing production, sponsorship, and dissemination of research by commercial enterprises and for-profit institutions or companies established by universities and their academic staff, the tendency to apply intellectual property rights and copyright to research and instructional materials, and raises in student charges. The more education and research are regarded as economic investments, the more their costs and returns are calculated according to market and proprietary principles. State subsidies have been reduced or removed in many countries, and student tuition rates have been raised to reflect the "real" costs of tertiary education.

It is not always compelling to attribute all these trends to contemporary globalization, but in the area of the new information and communication technologies (ICTs), the impact of globalization seems incontrovertible. The computerization of education involves the incorporation of ICTs into the knowledge activities of teaching, research, and publication. Much of the debate on globalization in higher education centers on the educational impact of ICTs; opinions differ sharply. The debate has centered on two issues: (a) the cost and profitability of online education; and (b) its pedagogical benefits. The jury is still out on both issues. Some studies assessing online education indicate that online programs are neither cheap to produce nor as profitable as originally anticipated. Some are even breaking, others are losing money, and a few are making money.

Some argue that we need to go beyond narrow financial calculations and the polar options of boosterism and rejectionism. Universities and academics, with the aim of harnessing the positive, and safeguarding their role as creators and certifiers of authoritative scholarship, must face the positive and negative potential of the new technologies squarely. Instead of wholesale embrace or dismissal, it is more productive to determine which technologies are useful for which students, for which subject matters, and for which purposes. Engagement with the new technologies allows universities to provide their students with critical technoliteracy, democratized and customized higher learning, and it also helps to shape the emerging ICT educational regime, for example, the tension between pedagogical and proprietary norms.

Since they are both repositories of information and media for knowledge production, the new technologies are not merely "delivery systems" that pass through colleges and universities, leaving their core values either unchanged or destroyed. Rather, the new technologies are an integral part of the complex and contradictory changes taking place in the conflicted terrain of higher education. If tapped carefully and creatively, they hold exciting possibilities for removing the spatiotemporal constraints that limit access for nontraditional students, promote student interaction and cooperative learning, pedagogical experimentation, collaborative research, and transnational exchanges. They can blur the distinctions between on-campus and off-campus teaching, between residential and distance education. In short, the impact of ICTs is ambiguous because, like all technologies, it is not simply an innocuous tool, rather it depends on its design and the technoculture it embodies and promises, the prevailing structural and institutional contexts in which it performs, and the broader material conditions and social relations in which it is articulated.

There has been increased connectivity of institutions that pertains to the increased emphasis on institutional cooperation and coordination within and across countries, a process facilitated by ICTs, competition from the new corporate interlopers of higher education, the rising costs of maintaining such expensive infrastructures as libraries, and pressures from students and for internationalization. International education cooperation involves a wide range of activities such as academic mobility, internationalization of curricula and programs, and networking and linking arrangements to research collaboration and joint publishing. In fact, Gibbons (1998, 2001) and his followers argue that networking and the shift of knowledge production to other sites is making disciplinary organization of universities and knowledge production obsolete, and encouraging transdiscplinary modes of organization or what, according to Gibbons and his collegues, is called Mode 2 knowledge production.[8]

Implications of Transformations on Academic Freedom and Knowledge Production

The developments identified above have complex and contradictory effects on academic freedom, by which I mean the autonomy of institutions and individuals within them to pursue knowledge production without undue pressure and prejudice. In other words, academic freedom is a functional condition, a philosophical proposition, and a moral imperative to the unfettered pursuit and dissemination of knowledge. Academic freedom allows universities to meet their responsibilities to society: speaking truth to power, promoting progress, and cultivating democratic citizenship. University autonomy, academic freedom, and social responsibility are essential for the production of the critical social knowledge that facilitates material and ethical advancement. In this context, the notion of social responsibility should not mean acquiescence to authoritarian regimes or repressive civil society institutions and practices. Rather, it requires a commitment to progressive social causes.[9]

It seems to me that a market-driven higher education system affects, indeed undermines, academic freedom in five major ways: (a) in terms of student access and solidarity; (b) disciplinary differentiation and devaluation; (c) integrity of research and publishing; (d) management and security of tenure; and (e) permeability and dilution of institutional traditions. The more education is regarded as an economic investment for individuals rather than a public good, the more its costs and returns are calculated according to market principles and hence student fees are raised, thereby making access more difficult for women, minority ethnic groups, and the rural poor. As fees rise becomes more differentiated across programs, learning increasingly becomes a market transaction and a consumer mentality takes hold among the high fee-paying students that can divorce them from those on subsidies, thereby weakening students' collective capacity to protect their rights and the quality of their education.

As learning becomes increasingly valued for its instrumentality, more emphasis is put on the technical and professional fields at the expense of the humanities and the basic sciences, and on applied research over basic research. This differentiation based on disciplinary marketability places faculty in the "unprofitable" disciplines at a grave disadvantage in institutional battles for resources, undermining their ability to undertake research and articulate a public voice. The devaluation of the humanities is increasingly evident in Africa's mushrooming private universities and the "privatised" public universities. The latter consist of "executive programs" for professionals studying part-time or full-fee courses targeted at full-time students. Many of these programmes focus on business studies and other marketable fields. The expansion in the number of private universities in Africa has been quite phenomenal. A few examples from East Africa will suffice. In Kenya, since the late 1950's the number of private universities grew from three to seventeen, so that there are now three times as many private universities as public ones (Ngome, 2003; Otieno, 2004). For its part, Uganda has sixteen universities, twelve of them private (Musisi, 2003). By 2003, the United Republic of Tanzania has seventeen universities, most of them private (Ministry of Science, Technology and Higher Education, 2004; Mkunde et al., 2003). The private universities, consisting of both religious and secular, profit and not-for-profit institutions tend to focus on religious training and business management, while the traditional humanities and basic sciences are underemphasized.[10]

The humanities find themselves increasingly challenged even in parts of the global North. Writing from Australia, Ang (1999) notes that "in the market-driven knowledge society of today, the usefulness of humanities knowledge is diffuse and inferential, and therefore difficult to measure and circumscribe." Thus,

> while the form of at least some humanities research may display Mode 2 characteristics, the difficulty to quantify and commodify its uses inhibits its

insertion into the emergent *social and economic arrangements* of Mode 2 knowledge production with their emphasis on calculable usefulness in clearly demarcated strategic contexts. The situation is not helped, by the fact that the professionalization of humanities scholarship has tended to move it more towards Mode 1 production practices, with its emphasis on disciplinary specialization and relative disconnection from the larger social world, rather than towards Mode 2, with its more problem-focused, multidisciplinary and collaborative orientation. This is the case also for cultural studies, as a brand of "applied humanities."

These sentiments are shared by Hunter (1999), from the United States, who observed,

[I]n the past five years or so—ever since the congressional debate that severely cut funding for the National Endowment for the Humanities and that left the National Endowment for the Arts hanging by a thread—just about everyone has become more concerned about how the public thinks of us and what we do. Even American humanists—with proper olfactory encouragement—can wake up and smell the coffee.

Davidson and Goldberg (2004) also lament:

Few observers of higher education would deny that support for the humanities is declining in an environment in which universities are increasingly ordered according to the material interests, conditions, and designs of the sciences, technology, and the professions. We contend, however, that if ever there were a time when society was in need of humanistic modes of inquiry, it is today. More than ever, we require the deep historical perspective and specialized knowledge of other cultures, regions, religions, and traditions provided by the humanities.

To save the field, they issued, "A Manifesto for the Humanities in a Technological Age," which calls for the development of new interdisciplinary paradigms, the creative integration of the new digital technologies, and the adoption of assertive strategies of advocacy.[11]

Clearly, notwithstanding all the fulminations about the so-called cultural wars and the self-aggrandizing posturing of the "posts"—poststructuralism, postmodernism, postcolonialism—there is concern among many humanities scholars in countries such as the United States that all is not well with the humanities, the foundational base of liberal education, in the brave new era of commercialized learning and commodified knowledge. In the United States voices from the humanities and the arts have been largely absent from public discourse addressing the fundamental questions of human existence—from war and violence, human rights, and the implications of the demographic transformations of states, nations, and regions, to the ethical and cultural implications of the defining technologies of our times (information technology, biotechnology, nanotechnology, and environmental technology), to the construction, reconstruction, and intersections of identities (social, religious, linguistic) at various scales from the local to the global. In a world of repetitive cable television news, what passes for public discourse in the United States is often nothing but mindless chatter by pompous, opinionated, and ignorant pundits.

The growth of commercially financed research is raising new concerns. Not only is "the constant search for funding time-consuming and inefficient," argues one author, corporate sponsors often seek to

> retain control of the direction of the research and even impose a new set of staff relationships. . . . The sponsor can also control intellectual property rights and even the right to publish the projected results. It can prevent the scientist from sharing research at an international conference and can even stop his or her work if the funder doesn't like the way it is going. The old expectation was that scientific expertise was global in its reach and exchange. . . . Today, the corporations are buying up this expertise, leaving very few voices to challenge what they are telling the world. (Evans, 2001, p. 17)

Stories abound of the research programs of entire centers or departments being mortgaged to corporations.[12] It has long been recognized among African academics that donor-funded research, including research sponsored by foundations of impeccable liberal credentials, often comes with thick strings attached that can compromise the choices and integrity of their research. This was discussed widely at the landmark conference on Academic Freedom convened by the Council for the Development of Social Science Research in Africa (CODESRIA) in Kampala, Uganda, in 1990.[13]

The emergence of powerful transnational academic publishers, who set exorbitant journal subscription prices that, effectively, bar access to information to all but those in rich institutions, is also a significant part of the intellectual property rights regime and corporate stranglehold on academic freedom and knowledge production. Resistance against such practices that undermine academic freedom is mounting. There are reported protests against blatant commercialization. Some universities have adopted conflict-of-interest guidelines; in September 2001 editors of the International Council of Medical Journal Editors adopted new rules concerning the ethics of clinical trial performance and reporting to prevent the publication and legitimation of dubious studies sponsored by pharmaceutical companies (Blumenstyk, 2000; Brainard, 2001; Kellogg, 2001); and reportedly in 2001, "more than 22,000 scientists from 161 countries launched a boycott of science publication editors and started campaigning for a 'public science library'" (Lefort, 2001, p. 24).

Business management models have given university administrators more executive powers, which has exacerbated management-faculty tensions and reduced the capacity of faculty to influence the running of their institutions. One result is that tenure is increasingly under threat where formal tenure systems are instituted. Once regarded as indispensable to the academic profession, and the pursuit of academic freedom, in many countries tenure, whatever it is called, is now widely perceived by hostile governments and the general public as an indefensible sinecure of lifelong employment, an entitlement that is as outdated and dangerous as the other "entitlements" being dismantled in the post-Fordist era of flexible production in the global North and postdevelopmentalist era of structural adjustment in the global South, and merciless free-market competition everywhere.

Universities in highly competitive academic systems, such as the United States, have responded by swelling the ranks of untenured part-time or adjunct faculty, who are crowded in introductory level courses, and to whom academic freedom is a myth. The proportion of part-time faculty in the country rose from 35 percent in 1991 to 44 percent in 2001 (The Chronicle of Higher Education, 2004, p. 28). To adjuncts, crowded in introductory level courses, academic freedom is a myth. "Adjuncts are getting dumped for things tenure-track scholars do with impunity—teaching controversial material, fighting grade changes, organizing unions. . . . All an institution has to do is not renew their contracts. No explanations required; no grievance procedures provided. Adjuncts just disappear" (Schneider, 1999, p. A18). Advocates of academic freedom in the United States warn that a dearth of academic freedom for almost half the professoriate threatens it for the other half.[14] Indeed, all is not well for the tenured half either; their academic freedom is often imperiled by the presence of speech codes and the absence of faculty unions on many campuses.

The erosion of the universities' old monopoly over knowledge production means that academics increasingly enjoy mobility between universities and other research sites outside universities, which offer them unprecedented opportunities to form networks, partnerships, and alliances that can not only enhance their research capacities but also protect them from the iniquitous tendencies of the academy. There can be little doubt that the proliferation of independent research centers and NGOs has saved many African academics from the penury and repression of their structurally adjusted universities. However, this institutional permeability also makes it harder to define academic freedom, to set its parameters in ways that are consistent with and that strengthen, rather than weaken, university traditions of academic freedom. Part of the widespread confusion over the meaning and implications of academic freedom arises from transformations and proliferation of sites of knowledge production spawned by the new cultural and political economies of knowledge production.

The implications of these transformations on gender are not only complex and contradictory but also quite varied in different regions and countries. On the one hand, corporatization reinforces authoritarian and masculinized management styles of higher educational institutions. On the other hand, as the access of women to universities increases, long-entrenched androcentric practices and perspectives come under scrutiny and challenge. Moreover, the flexibilities of lifelong learning appear more accommodating to women's occupational life cycles and experiences and demand reorientations from many men accustomed to more uninterrupted occupational lives. At the same time, however, the marketization of universities requires academics to work longer hours. This leads not only to higher stress levels but also reinforces old gender differentiations between men and women based on the unequal domestic division of labor for those with family responsibilities. Furthermore, as the business practice and ideology of flexible production infiltrate the universities, the academy becomes increasingly divided between an elite professoriate with all the privileges of academia, including higher salaries and benefits, and a growing mass of the lumpen professoriate of part-time, poorly paid academics, among which women tend to predominate.

Thus, at the very time that feminist scholarship is expanding as a result of the entry of more women in the academy, humanistic knowledge as a whole, in which feminist paradigms, pedagogy, and praxis are lodged, become devalued. In response, many humanities scholars, including feminists, have sought solace in either more theoretical rigor or applied research. The search for rigor, often spawned by an awareness of being threatened by the "hard" sciences, is evident in the rise of the turgid and difficult rhetoric of postmodernism and postcolonialism, especially in the academies of the global North. This has often turned discourses in the humanities, including many feminist writings, into self-referential conversations incomprehensible to the public constituencies they once professed (and sometimes still profess) to speak for. Elsewhere, especially in the global South, unease with theoretical fetishism and fidelity to the unyielding dreams of development, not to mention the pecuniary demands for survival, lead to the romanticizing of relevance, whose consummation is sought, not in social movements, but in consultancies for NGOs and international donor agencies.

In many African countries the percentage of female faculty at the end of the 1990s was lower than that of female students, ranging from as low as 6.1 percent in Ethiopia to 19.7 percent in Uganda to 35 percent in South Africa (Ebrima, 2000; Kwesiga, 2002; Meena, 2001). At South African universities the bulk of women faculty are at the rank of lecturer and below. In 2002, women made up 55 percent of those at the rank of junior and others (up from 53 percent in 1995), 53 percent of lecturers (up from 46 percent), 38 percent of senior lecturers (up from 28 percent), and 19 percent of professors (up from 13 percent) (Balintulo, 2004; Council for Higher Education, 2004, p. 79–80). At Uganda's venerable Makerere University, in 1998–1999, there were only 2 women professors out of 47 (compared to 2 out of 39 in 1996–1967), 3 associate professors out of 54 (as compared to 2 out of 49), 49 senior lecturers out of 202 (compared to 21 out of 180), 104 lecturers out of 441 (compared to 108 out 442), 3 assistant lecturers out of 153 (compared to 27 out of 150) (Musisi and Muwanga, 2003). At the University of Dar-es-Salaam in the United Republic of Tanzania, where women made up 11 percent of the faculty in 1999–2000, only 4 out of the 40 full professors in 1996 were women (Yahya-Othman, 2000, p. 39).

The gender gap is of course not confined to Africa. In the United States, women comprised 38.4 percent of all full-time faculty members, but only 22.7 percent of professors, and 36.8 percent of associate professors, 44.8 percent of assistant professors, 50.6 percent of instructors, 53.3 percent of lecturers, and 44.9 percent of others. The rates of underrepresentation in the higher ranks and overrepresentation in the lower ranks tend to be even more pronounced for minority women, not to mention minority men, especially black and Hispanic (The Chronicle of Higher Education, 2004, p. 29). In Canada, women made up 30 percent of faculty in 2003, up from 20 percent in 1990, but they comprised only 17 percent of professors, up from 8 percent in 1990 (Statistics Canada, 2005).

Structural Adjustment and African Universities

Like universities elsewhere, during the past two decades African universities have been undergoing unprecedented change and confronting multiple challenges, both old and new. But these challenges have had a particular configuration and intensity. Struggles of various kinds and force have been and are being waged within and outside the university system, on the contemporary interpretation and operationalization of its mission. State, economy, and society are all in a state of flux, with consequences for popular perceptions of the role, place, and relevance of the university in personal and societal progress.

Central to these challenges has been the weakened capacity of the state in most African countries after nearly two decades of unrelenting economic crises and orthodox structural adjustment, major shifts in the composition and orientation of the student body, changes in the content and system of instruction, and the systematic brain drain that has robbed the university community of some of its most talented members. According to some estimates, an average 20,000 highly educated Africans have been migrating to the global North every year since 1990.[15] In addition, growing numbers of Africans trained abroad are not returning home. A study conducted by the United States Social Science Research Council on the rates of return of African PhDs trained in North America between 1986 and 1996 (Pires et al., 1999) showed that more than one-third, or 36 percent to be exact, stayed in North America, and 2 percent went to Europe or elsewhere, while 57 percent returned to their countries of origin, and a further 5 percent to other African countries. All this has led to an interesting spectacle in which African immigrants now constitute the most educated population of any group in the United States, native born or foreign born. According to the United States 2000 Census, there were 700,000 African-born residents in the United States. Among the African-born residents, aged 25 and over, 49.3 percent had a bachelor's degree or more, as compared to 25.6 percent for the native-born population and 25.8 percent for the foreign-born population as a whole (U.S. Census Bureau, 2001, pp. 36–37). Among these, highly educated African immigrants are to be found large numbers of academics who have left Africa's beleaguered universities (Zeleza, 2004).

The historic assumptions that informed the founding of the modern African university, and which shaped broad social responses to it, have all but evaporated in the face both of the crises of the postindependence nationalist project, and the neoliberal onslaught on the entire fabric of the postcolonial model of development—a model in which the university occupied a central, multifaceted role. It may seem like distant memory already that as late as the 1990s, powerful international forces promoting a neoliberal agenda and led in the African context by the World Bank (1988, 1994a, 1994b) had suggested, literally, that Africa had no need for universities because the return on investment that it received from its expenditure outlay was both too low and unjustifiable. Rather than establish, maintain, and invest in the university, it was argued that Africa would be better served by investing in primary education and vocational education, while exploring more cost-effective foreign options for university-level training, that it would probably be cheaper, more cost effective and beneficial to train African students in universities abroad. The serious controversy that the World Bank position generated and the contestations it produced on campuses across the continent constituted an important element of the politics of university reform in Africa throughout the 1990s.

More recent World Bank publications and pronouncements suggest a radical rethinking of its antiuniversity orientation of the 1990s, even if this has neither gone hand in hand with an admission that its earlier position, promoted with all its donor might, was misplaced nor an acceptance of responsibility for some of the damage that the university system suffered as a consequence of its determinant policy influences on African governments. Today, the World Bank asserts the view as to the important place of the higher education system, with the university at its core, in the prospects for development. While this U-turn may be welcome for all it is worth, it still bears emphasizing that the strong market-instrumentalist logic permeating the

World Bank's new approach poses fresh challenges, which require to be addressed in thinking of a vision and a role for the African university (World Bank and UNESCO, 2000). The issues that require careful balancing are many indeed: autonomy and viability, expansion and excellence, equity and efficiency, access and quality, and authority and accountability. Additionally, there are questions of representation and responsibility, diversification and differentiation, internationalization and indigenization, and global presence/visibility and local anchorage. No less crucial are the challenges of academic freedom and professional ethics, privatization and the public purpose, teaching and research, community service/social responsibility and consultancy, and diversity and uniformity. There are also concerns about the preservation of local knowledge systems and the adoption of global knowledge systems, knowledge production and knowledge dissemination, and the knowledge economy and the knowledge society.

The contexts and content of the challenges redefining the university and securing its place of course differ among countries, even within Africa, but they all reflect the decomposition of the old social contract between the university, the state, and society in which higher education was valued as a public and intellectual good that, moreover, dovetailed into visions of nation building and national development. As market imperatives and ideology have gained or are struggling to gain supremacy, universities are increasingly valorized or find themselves compelled to seek valorization for their private and vocational good. Several issues are particularly important to highlight. First, the implications of the new funding strategies that the cash-strapped African governments working under IMF/World Bank conditionality are being forced to impose on the universities. Second, the expansion of private, including religious universities, poses new regulatory challenges with regard to quality in various spheres and generating debates about access, equity, diversity, and the secular foundations of the higher education system as a whole. Third, the pressures associated with the process of massification in the higher education system pose challenges for both the development and delivery of academic programs and the governance of the university. Fourth, the growth in the trade in educational services, which the World Trade Organization's General Agreement on Trade in Services (GATS) seeks to regulate and encourage, raises important questions about the viability and global competitiveness of African universities.[16]

In the face of the global and local changes that are occurring, and the challenges that need to be faced, the questions that are posed as the struggle for the African university continues are numerous: How are these changes affecting the teaching and research systems in African universities? What does globalization specifically mean for these universities? What have been the policy responses in terms of liberalization and privatization of the higher educational sector, and the effects in terms of academic exchanges within the continent itself and between African countries and other parts of the world? Does the concept of the public university have a future at all in Africa? Is there a place for a socially responsible and responsive African university in the twenty-first century? How can African universities deal with the brain drain and make better use of the African intellectual diaspora to assist in the Africanization of global scholarship and globalizing of African scholarship? What is the impact of the changing technological environment on African universities? Do the universities have a role to play in the preservation of local/national identities in the face of onslaughts from powerful external forces and influences, and without prejudice to any desire they might have to be established as veritable centers of excellence comparable to any other elsewhere in the world? How are they handling the issues of access and equity with reference to male and female participation rates, and the knotty problems of ethnicity, class, religion, and for some such as South Africa, race, as well as the articulation of university education with primary and secondary education? How can their internal management and governance structures be improved? What has been the role of governments, donors, student, and academic staff associations in the development of university administrative systems and cultures? What is the nature of linkages between the universities and various economic sectors and the labor market, including industry, agriculture, services, and the public sector? How would reforms in higher education enhance its contribution to the promotion of sustainable human development? What new tasks should African institutions of higher education undertake to meet the changing economic needs of African economies? How, indeed, can the universities best serve their societies as a whole, in addressing pressing social issues from the HIV/AIDS pandemic to civil conflicts at the same time as they seek to protect and promote their own institutional and intellectual autonomy?

Conclusion

Many of these questions are examined in considerable detail in the two-volume study, African Universities in the Twenty-First Century, which I coedited (Zeleza and Olukoshi, 2004a, 2004b). In fact, an enormous amount of research is now being conducted on these questions and many others not listed here. In addition to the benefits that research on the African higher education system confers on the quest for sustainable reforms with a progressive, transformatory edge in the university, such an exercise is also indispensable to the building of a body of knowledge generated from within Africa on the budding field of higher education studies. The field of African higher education research has emerged as a critical area of inquiry, part of the strategic rethinking, reorientation, and revitalization of African universities. For it to be comprehensive in its coverage of the issues at the center of the problems and prospects of the African university, the research that needs to be undertaken will necessarily have to be multidisciplinary in orientation. Fortunately, research that is going on in African higher education today and that is leading to a corpus of works on the subject is being conducted from a variety of disciplinary angles.

An agenda for research on the African context will need to both grapple with global trends in higher education in terms of what they mean for Africa and its universities, and also capture the specific needs that arise from the continent's own experiences. In my opinion, such an agenda needs to be framed around five broad sets of issues. First, there is need to examine systematically the philosophical foundations of African universities. Included in this context are issues pertaining to the principles underpinning public higher education in an era of privatization, the conception, content, and consequences of the reforms currently being undertaken across the continent, and the public-private interface in African higher education systems.

The second set of issues center on management, how African universities are grappling with the challenges of quality control, funding, governance, and management in response to the establishment of new regulatory regimes, growing pressures for finding alternative sources of funding, changing demographics and massification, increasing demands for access and equity for underrepresented groups including women, and the emergence of new forms of student and faculty politics in the face of democratization in the wider society. Third, there are pedagogical and paradigmatic issues, ranging from the languages of tuition in African universities and educational systems as a whole to the dynamics of knowledge production—the societal relevance of the knowledges produced in African higher education systems, and how these knowledges are disseminated and consumed by students, scholarly communities, and the wider public.

Fourth, the role of universities in the pursuit of the historic project of Africa nationalism: decolonization, development, democratization, nation building, and regional integration. African universities have served as important ideological and institutional instruments in pursuit of some of these objectives and, in this regard, there is need to investigate contemporary trends and their likely trajectories, in the uneven and changing relations among universities and the state, civil society, and industry in different countries and regions, as well as the role of universities in helping to manage and resolve the various crises that confront the African continent from civil conflicts to disease epidemics including HIV/AIDS. Also, the part that universities have played, and can play in the future, to promote or undermine the Pan-African Project is a of great interest as African states, through the African Union, renew their efforts to achieve closer integration within Africa and between Africa and its diasporas.[17]

Finally, there is the question of globalization, the impact of trends associated with the new ICTs, the expansion of transborder or transnational provision of higher education, and trade in educational services under the GATS regime. Critical in this context for Africa is the changing role of external donors from the philanthropic foundations to the World Bank and other international financial institutions and multilateral agencies. The impact of these trends on African higher education and vice versa is of utmost importance and provides one area of fruitful collaboration between researchers from Africa and other world regions.

As noted earlier, a considerable amount of research is already being conducted on the various challenges that confront African universities. In addition to the work being undertaken by individual researchers and institutions in specific countries, there is need for more comprehensive projects that are collaborative and comparative in scope on the fives sets of issues identified above. There are numerous agencies and organizations that have been set up in recent years to promote the revitalization of African universities.[18] The Association of African Universities (AAU) is best placed as the organization that can coordinate such projects. The AAU can also facilitate research cooperation with regional and international partners engaged in coordinating research on transformations in higher education in different parts of the world.

Let me conclude by saying that the need to redefine and defend the role of higher education institutions as important centers for the production of critical social knowledge has never been greater than it is now. The challenge is to ensure that marketization does not turn higher educational institutions into vocational schools and consultancy outfits and that, as they transform themselves, they remain committed to the production of knowledge for social progress rather than the peddling of information for private profit. It is imperative that the global South, including Africa, be centrally involved in discourses and debates about the dynamics and directionality of globalization in general, and in mapping the meanings, content, and implications of the developmentalist and discursive fads of our age—knowledge society and knowledge economy—before they acquire the coercive conceit of immutability that characterized neoliberalism in the 1980s and 1990s that was exported to the global South, in the form of structural adjustment programs, with missionary zealotry by the gendarmes of global capitalism.

Notes

This essay was originally written for the endnote address at the Global Research Seminar on Knowledge Society versus Knowledge Economy: Knowledge, Power and Politics, Organized by UNESCO's Forum on Higher Education, Research and Knowledge, Paris, December 8–9, 2003. It has been revised to reflect comments from colleagues, especially Cassandra R. Veney, and one of the editors of this volume, Sverker Sörlin.

1. The literature on globalization is now vast. For a summary of this literature and my own understanding of the globalization debates, see Zeleza (2003a).
2. For a more detailed analysis of views by leading African intellectuals on globalization and Africa's engagements with the rest of the world since the sixteenth century, see Magubane and Zeleza (2003). A flavor of African perspectives on globalization can also be found in the collections by Nabudere (2000) and Aina et al. (2004).
3. The literature on knowledge economies and knowledge societies is also vast and growing rapidly. For useful succinct discussions of these debates and implications for higher education research, see de Weert (1999), Enders (1999), and Blackmore (2002). The term "expertoisie" is used by Hodges and Lustig (2002). For longer analyses see Stehr (1994), Barnett and Griffin (1997).
4. The literature on hegemony is also extensive. For some illustrative analyses see Chase-Dunn et al. (1994), Mitchell (1990), Lears (1985), Bates (1975), Joseph (2002), Banerjee (2001), Ludden (2001), and Lem and Leach (2002).
5. In the United States, for example, the percentage share of state appropriations for public higher education funding fell from 45.6 in 1980–1981 to 35.6 in 2000–2001. In the meantime, public higher education continued to expand more rapidly than private higher education, as it had done since 1950. Enrollments in public higher education rose from 9.5 million in 1980 to 12.2 million in 2001, while enrollment in private higher education rose from 2.5 to 3.7 million during the same period. To offset diminishing state support, tuition, and fees have risen sharply from 12.9 percent of the total income of public higher education in 1980 to 18.1 percent in 2001. See National Center for Education Statistics, Postsecondary Education, p. 222 at http://nces.ed.gov/pubs2005/2005025c1.pdf and p. 388 at http://nces.ed.gov/pubs2005/2005025c3.pdf (retrieved April 12, 2005). For a detailed account of the decline in government expenditure for higher education in African countries, see Woodhall (2003). Tertiary expenditure per student as a percentage of GNP per capita declined in most world regions between 1980 and 1995, from 39 to 26 among the high-income countries, and from 259 to 91 among the low- and middle-income countries (for sub-Saharan Africa the decline was from 802 to 422, East Asia and Pacific from 149 to 76, South Asia from 143 to 74, Europe and Central Asia from 67 to 36, and Middle East and North Africa from 194 to 82; Latin America and the Caribbean were the exceptions, where it rose from 19 to 143) (Woodhall, 2003, p. 46).
6. According to López-Bassols (1998), "Government-funded R&D has been declining in the large OECD countries. It averaged between 0.6 per cent and 1.0 per cent of GDP in 1995, with its annual growth rate in the OECD area as a whole decreasing in real terms from 4.7 per cent during 1981–85, to a drop of 0.4 per cent in 1991–95.... The situation for universities has changed, too: the share of university research in R&D expenditure increased slowly in most OECD countries in the late 1980s but levelled out, and even declined in some countries; in Canada, for example, it fell from more than 26 per cent of national R&D in 1991 to less than 22 per cent five years later. In all OECD countries, the bulk of university research funding comes from the public sector. Lower government funding has posed a serious challenge for higher education. Universities have had to improve their efficiency, find alternative sources of funding and increase co-operation with business as well as with other universities.... There is nevertheless some concern that the decline in funding will affect future contributions to the body of knowledge by undermining the ability of universities to fulfil their vital role in carrying out basic, long-term research."
7. The literature on globalization and universities is also large. The articles mentioned in note 4 above provide useful summaries. See also Buchbinder (1993), Gibbons et al. (1994), Clark (1995), Burbules and Callister (2000), Brooks and Mackinnon (2001), and Marginson and Considine (2001).

8. For a nuanced critique of the Gibbons thesis in the South African context, see Subotzky and Cele (2004). They explain the political, policy, and paradigmatic contexts in which the Gibbons thesis was initially appropriated in South Africa, and the subsequent critical responses to it, in which critics questioned its accuracy in describing the purported changes taking place in higher education and their efficacy. From their own survey of "strategic" collaborative projects at four institutions, Subotzky and Cele demonstrate that the Gibbons model is too narrow and dichotomous to account for the broader and heterogeneous range of knowledge production activities being generated out of the interactions between the academy, the corporate sector, government, NGOs, and donors, in which disciplinary knowledges are not so much as being replaced, indeed cannot be discarded in teaching without disastrous consequences, but are being combined and complemented with predisciplinary and interdisciplinary knowledges in complex and dynamic ways. They warn against the literal interpretation and employment of interdisciplinary and applications-driven research and curricula models, which can be counterproductive in the absence of foundational disciplinary knowledges.
9. The literature on academic freedom is uneven. For a more comprehensive discussion of the effects of the neoliberal regime on academic freedom, see my article (Zeleza, 2003b). For African discourses, see Diouf and Mamdani (1994). Also see Singh (2001) Bloom (2001), and the World Bank and UNESCO (2000).
10. Nevertheless, in all the three East African countries the vast majority of students are enrolled in the public universities. For example, in Kenya in 1999–2000 private universities accounted for only 14.2 percent of the total 49,400 undergraduate student population, up from 12.8 percent (out of the 40,816 undergraduate students enrolled in 1996–1997). In Tanzania the private universities accounted for 16.1 percent of the undergraduate student population of 9,054. For detailed data on the number and structure of universities in each African country from Algeria to Zimbabwe, see the comprehensive compendium by Teferra and Altbach (2004).
11. For responses to the "Manifesto" see Letters to the Editor: To Save the Humanities, We Need to Do More Than Talk, The Chronicle of Higher Education, 50(30), p. B4. A sense that the humanities are becoming marginalized in the United States was apparently pervasive at the 2005 Annual Meeting of the Modern Language Association as reported by Richard Bryne (2005).
12. See, for examples, stories from the University of Toronto (Turk, 2001) and the University of California at Berkeley (Elliot, 2001).
13. The conference led to the adoption of the "Kampala Declaration on Intellectual Freedom and Social Responsibility." The declaration is included as an appendix in Diouf and Mamdani's (1994) edited collection of the conference proceedings. For a fascinating commentary on this issue by the then regional representative of the Rockefeller Foundation, see Court (1990). For a brilliant satire of the pitfalls of donor-funded research, see Hirji (1990). Also see the biting critiques by Mkandawire (1989, 1997b), once the executive secretary of CODESRIA, and since 1997 the executive director of the United Nations Research Institute for Social Development. I have written extensively on this elsewhere, see Zeleza (1997, especially chapters 3 and 4).
14. Roger Bowen the head of the American Association of University Professors (AAUP) "calls the 'adjunctification' of the faculty one of the top two or three problems facing all of higher education. With half of the faculty now made up of part-timers, academe is moving toward a piecework system similar to that of farm laborers," see Smallwood (2004). The following recent reports provide a broad overview of the controversies raging in the American academy concerning the effects of "adjunctification" on academic freedom: Bronfenbrenner and Juravich (2001), Murphy (2002), Wilson (2002), Smallwood (2003), and Fogg (2001, 2004a, 2004b).
15. See the special issue of African Issues, on Africa's brain drain to the global North, that I coedited (Zeleza and Veney, 2002).

16. For a detailed analysis of the implications of GATS on African higher education see my keynote address delivered at the 11th General Conference of the Association of African Universities (Zeleza, 2005).
17. The African Union convened an important conference of intellectuals to devise new strategies for Pan-Africanism in the new century. See, African Union, Conference of Intellectuals from African and the Diaspora, Africa in the 21st Century: Integration and Renaissance, Dakar, Senegal, October 7–9, 2004, at http://www.au-ciad.org/.
18. For a list of the key international and national organizations involved in African higher education research, funding, and advocacy, see International Education Resources: Africa, at http://www.wes.org/ewenr/ResearchAfrica.htm.

References and Works Consulted

Aina, T. (1997) *Globalization and Social Policy in Africa: Issues and Research Direction*. Dakar: Codesria Monograph Series.

Aina, T. A., S. L. Chachage, and E. Annan-Yao (eds.) (2004) *Globalization and Social Policy in Africa*. Dakar: Codesria Book Series.

Ake, C. (1995) The New World Order: A View from Africa. In H. Hans Henrik and G. Sørensen (eds.) *Whose World Order: Uneven Globalization and the End of the Cold War*. Boulder, CO: Westview, pp. 19–42.

Amin, S. (1997) *Capitalism in the Age of Globalization*. London: Zed Books.

Ang, I. (1999) Who Needs Cultural Research? http://www.fas.harvard.edu/~chci/angfv.html (accessed April 10, 2005).

Balintulo, M. (2004) The Role of the State in the Transformation of South African Higher Education (1894–2002): Equity and Redress Revisited. In P. T. Zeleza and A. Olukoshi (eds.) *African Universities in the Twenty-First Century*. Vol. II. Knowledge and Society. Dakar: Codesria Book Series, pp. 441–458.

Banerjee, H. (2001) *Inventing Subjects: Studies in Hegemony, Patriarchy, and Colonialism*. New Delhi: Tulika.

Barnett, R., and A. Griffin (1997) *The End of Knowledge in Higher Education*. Trowbridge: Redwood Books.

Bates, T. R. (1975) Gramsci and the Theory of Hegemony. *Journal of the History of Ideas* 36(2), pp. 351–366.

Blackmore, J. (2002) Globalization and the Restructuring of Higher Education for New Knowledge Economies: New Dangers or Old Habits Troubling Gender Equity Work in Universities? *Higher Education Quarterly* 56(4), pp. 419–441.

Bloom, D. (2001) Higher Education in Developing Countries: Peril and Promise. *SRHE International News* 46, pp. 20–24.

Blumenstyk, G. (2000) A New Website Details the Corporate Ties of Some Researchers. *The Chronicle of Higher Education* 47(38), p. A25.

Brainard, J. (2000) At NIH Meeting, Scientists Debate When They Should Reveal Financial Interests to Volunteers. *The Chronicle of Higher Education* March 30, p. A43.

Bronfenbrenner, K., and T. Juravich (2001) Universities Should Cease Hostilities with Unions. *The Chronicle of Higher Education* January 19, p. B24.

Brooks, A., and A. Mackinnon (eds.) (2001) *Gender and the Restructured Universities*. Buckingham: Open University Press.

Bryne, R. (2005) Scholars Mull their Separation from the Mainstream. *The Chronicle of Higher Education* 51(18), p. A31.

Buchbinder, H. (1993) The Market Oriented University and the Changing Role of Knowledge. *Higher Education* 26(3), pp. 331–347.

Burbules, N. C., and T. A. Callister (2000) Universities in Transition: The Promise and the Challenge of New Technologies. *Teachers College Record* 102(2), pp. 271–293.

Chase-Dunn et al. (1994) Hegemony and Social Change. *Mershon International Studies Review* 38(2), pp. 361–376.

Clark, B. R. (1995) *Creating Entrepreneurial Universities: Organizational Pathways of Transformation*. Oxford: IAU Press and Pergamon.

Council for Higher Education (2004) *South African Higher Education in the First Decade of Democracy*. Pretoria: Council for Higher Education.

Court, D. (1990) Universities and Academic Freedom in East Africa, 1963–1983: Random Reflections from a Donor Perspective. Paper presented at the CODESRIA Symposium on Academic Freedom, Research and the Social Responsibility of the Intellectual in Africa, Kampala, Uganda, November.

Davidson, C., and D. T. Goldberg (2004) A Manifesto for the Humanities in a Technological Age. *The Chronicle of Higher Education* 50(23), p. B7.

de Weert, E. (1999) Contours of the Emergent Knowledge Society: Theoretical Debate and Implications for Higher Education Research. *Higher Education* 38, pp. 49–69.

Diouf, M., and M. Mamdani (eds.) (1994) *Academic Freedom in Africa*. Dakar: Codesria.

Ebrima, E. (ed.) (2000) *Women in Academia: Gender and Academic Freedom in Africa*. Dakar: Codesria Book Series.

Elliot, V. (2001) Who Calls the Tune? *The UNESCO Courier* November, pp. 21–22.

Enders, J. (1999) Crisis? What Crisis? The Academic Profession in the Knowledge Society? *Higher Education* 38, pp. 71–81.

Evans, G. (2001) Leaving Room for Dissent. *The UNESCO Courier* November, p. 17.

Fogg, P. (2001) U.S. Court Lets Adjunct Professor Sue Officials Who Refused to Renew His Contract. *The Chronicle of Higher Education*, p. A18.

—— (2004a) Faculty Union Calls for Better Treatment of Graduate Assistants. *The Chronicle of Higher Education* 51(9), p. A16.

—— (2004b) For these Professors, "Practice is Perfect." *The Chronicle of Higher Education* 50(32), p. A12.

Gibbons, M. (1998) *Higher Education Relevance in the 21st Century*. Washington, DC: World Bank.

—— (2001) Globalization in Higher Education: A View for the South. *SRHE International News* 46, November, pp. 4–12.

Gibbons, M., C. Limoges, H. Nowotny, S. Schwartzman, P. Scott, and M. Trow (1994) *The New Production of Knowledge: The Dynamics of Science and Research in Contemporary Societies*. London: Sage.

Hirji, K. F. (1990) Academic Pursuits Under the Link. *Codesria Bulletin* 1, pp. 9–16.

Hodges, D. C., and L. Lustig (2002) Bourgeoisie Out, Expertoisie In The New Political Economies at Loggerheads. *The American Journal of Economics and Sociology* 61(1), pp. 367–381.

Hunter, J. P. (1999) Going Downtown: The Necessity—and Hazards—of Telling the Public about Humanities Research, http://www.fas.harvard.edu/~chci/hunterfv.html (accessed April 10, 2005).

Joseph, J. (2002) *Hegemony: A Realist Analysis*. New York: Routledge.

Kellogg, A.P. (2001) Leading Medical Journals Hope to Curb Drug Companies' Publishing Influence. *The Chronicle of Higher Education* August 6.

Kwesiga, J.C. (2002) *Women's Access to Higher Education in Africa: Uganda's Experience*. Kampala: Fountain Publishers.

Lears, J. T. J. (1985) The Concept of Hegemony: Problems and Possibilities. *American Historical Review* 90(3), pp. 567–593.

Lefort, R. (2001) Barbed Wire in the Research Field. *The UNESCO Courier* November, pp. 24–25.

Lem, W., and B. Leach (eds.) (2002) *Culture, Economy, Power: Anthropology as Critique, Anthropology as Praxis*. Albany: State University of New York Press.

López-Bassols, V. (1998) How R&D is Changing. *OECD Observer* (retrieved April 10, 2005), no. 213, at http://www1.oecd.org/publications/observer/213/Article4_eng.htm.

Ludden, D. (2001) *Reading Subaltern Studies: Critical History, Contested Meaning, and the Globalisation of South Asia*. Delhi: Permanent Black.

Magubane, Z., and P. T. Zeleza (2003) Globalization, Intellectuals, and Africa. In M. Steger (ed.) *Rethinking Globalism*. New York: Rowman and Littlefield, pp. 165–177.

Marginson, S., and M. Considine (2001) *The Enterprise University: Power, Governance and Reinvention in Australia*. Melbourne: Cambridge University Press.

Meena, R. (2001) Women's Participation in Higher Levels of Learning in Africa: Interventions to Promote Gender Equity. Paper presented at the Carter Lectures, Government and Higher Education, University of Florida, Gainesville, March 22–25.

Ministry of Science, Technology and Higher Education (2004) Report to the Public on the Probe Team on Students Crises in Higher Education Institutions in Tanzania, http://www.tanzania.go.tz/education.html (accessed April 12, 2005).

Mitchell, T. (1990) Everyday Metaphors of Power. *Theory and Society* 19(5), pp. 545–577.

Mkandawire, T. (1989) Problems and Prospects of the Social Sciences in Africa. *Eastern Africa Social Science Review* 5, pp. 1–17.

—— (1997a) Globalization and Africa's Unfinished Agenda. *Macalaster International* 7, pp. 71–107.

—— (1997b) The Social Sciences in Africa: Breaking Local Barriers and Negotiating International Presence. *African Studies Review* 40(2), pp. 15–36.

—— (1998) Thinking About Development in Africa. Study No. 9, African Development in a Comparative Perspective. Geneva, United Nations Conference on Trade and Development.

Mkandawire, T., and C. C. Soludo (1999) *Our Continent, Our Future: African Perspectives on Structural Adjustment*. Trenton, NJ: Africa World Press.

Mkandawire, T., and C. C. Soludo (eds.) (2003) *African Voices on Structural Adjustment: A Companion to Our Continent, Our Future*. Trenton, NJ: Africa World Press.

Mkunde, D., B. Cooksey, and L. Levey (2003) *Higher Education in Tanzania: A Case Study*. Oxford: James Currey.

Murphy, M. (2002) Adjuncts Should Not Just be Visitors in the Academic Promised Land. *The Chronicle of Higher Education* March 29, p. B14.

Musisi, N. (2003) Uganda. In D. Teferra and P. G. Altbach (eds.) *African Higher Education: An International Reference Handbook*. Bloomington and Indianapolis: Indiana University Press, pp. 611–623.

Musisi, N., and N. K. Muwanga (2003) *Makerere University in Transition 1993–2000*. Oxford: James Currey.

Nabudere, D. W. (2000) Globalization, the African Post-Colonial State, Post-Traditionalism, and the New World Order. In D. W. Nabudere (ed.) *Globalization and the Post-Colonial African State*. Harare: Sapes Books, pp. 7–55.

—— (ed.) (2000) *Globalization and the Post-Colonial African State*. Harare: Sapes Books.

Ngome, C. (2003) Kenya. In D. Teferra and P. G. Altbach (eds.) *African Higher Education: An International Reference Handbook*. Bloomington and Indianapolis: Indiana University Press, pp. 359–371.

Otieno, W. (2004) The Privatization of Kenyan Public Universities. *International Higher Education* Summer, pp. 13–14.

Petras, J. (1999) Globalization: A Critical Analysis. *Journal of Contemporary Asia* 29(1), pp. 3–37.

Pires, M., R. Kassimir, and M. Brhane (1999) *Investing in Return: Rates of Return of African Ph.D.s Trained in North America*. New York: Social Science Research Council.

Schneider, A. (1999) To Many Adjunct Professors, Academic Freedom Is a Myth. *The Chronicle of Higher Education* December 10, p. A19.
Singh, M. (2001) Re-Inserting the "Public Good" into Higher Education Transformation. *SRHE International News* 46, pp. 24–27.
Smallwood, S. (2003) Union Without a Contract. *The Chronicle of Higher Education* 50(13), p. A8.
—— (2004) A Force of Nature Takes Over at the AAUP. *The Chronicle of Higher Education* 50(40), p. A8.
Statistics Canada (2005) The Rising Profile of Women Academics. *Perspectives on Labor and Income* 6(2), pp. 75–101. Available at http://www.statcan.ca/english/studies/75–001/peonline.htm.
Stehr, N. (1994) *Knowledge Societies*. London: Sage Publications.
Subotzky, G., and G. Cele (2004) New Modes of Knowledge Production: Peril or Promise for Developing Countries? In P. T. Zeleza and A. Olukoshi (eds.) *African Universities in the Twenty-First Century*. Vol. II. Knowledge and Society. Dakar: Codesria Book Series, pp. 341–362.
Teferra, D., and P. G. Altbach (eds.) (2004) *African Higher Education: An International Reference Handbook*. Bloomington and Indianapolis: Indiana University Press.
The Chronicle of Higher Education (2004) Almanac Issue 2004–2005, Li, 1 (August 27).
Turk, J. L. (2001) Anatomy of a Corporate Takeover. *The UNESCO Courier* November, pp. 18–20.
U.S. Census Bureau (2001) *Profile of the Foreign Born Population in the United States*. Washington, DC: U.S. Census Bureau.
Wilson, R. (2002) Working Half Time on the Tenure Track. *The Chronicle of Higher Education* January 25, p. A10.
Woodhall, M. (2003) Financing and Economics of Higher Education in Africa. In D. Teferra and P. G. Altbach (eds.) *African Higher Education: An International Reference Handbook*. Bloomington and Indianapolis: Indiana University Press, pp. 44–52.
World Bank (1988) *Education in Sub-Saharan Africa: Policies for Adjustment, Revitalisation and Expansion*. Washington, DC: World Bank.
—— (1994a) *World Bank's Role in Human Resource Development in Sub-Saharan Africa*. Washington, DC: World Bank.
—— (1994b) *Higher Education: The Lessons of Experience*. Washington, DC: World Bank.
World Bank and UNESCO (2000) *Higher Education in Developing Countries. Peril and Promise*. Washington, DC: World Bank.
Zeleza, P. T. (1997) *Manufacturing African Studies and Crises*. Dakar: Codesria Book Series.
—— (2003a) *Rethinking Africa's Globalization*. Vol. 1. The Intellectual Challenges. Trenton, NJ: Africa World Press.
—— (2003b) Academic Freedom in the Neo-Liberal Order: Governments, Globalization, Governance, and Gender. *Journal of Higher Education in Africa* 1(1), pp. 149–194.
—— (2004) The African Academic Diaspora in the United States and Africa: The Challenges of Productive Engagement. *Comparative Studies of South Asia, Africa, and the Middle East* 24(1), pp. 265–278.
—— (2005) Transnational Education and African Universities. Paper presented at the 11th General Conference of the Association of African Universities, "Transnational Education and African Universities." Cape Town, South Africa, February 21–25.
Zeleza, P. T., and A. Olukoshi (2004a) *African Universities in the Twenty-First Century*. Vol. 1. Liberalization and Internationalization. Dakar: Codesria Book Series.
—— (2004b) *African Universities in the Twenty-First Century*. Vol. 2. Knowledge and Society. Dakar: Codesria Book Series.
Zeleza, P. T., and C. R. Veney (eds.) (2002) *African Issues* xxx(1), special issue on "The African Brain Drain to the North: Pitfalls and Possibilities."

Chapter Four
Knowledge and Equity: Unequal Access to Education, Academic Success and Employment Opportunities: The Gender Balance in Algeria

Nouria Benghabrit-Remaoun

Introduction

Upon gaining independence a country always perceives education as an essential tool toward development, and one of the major aims of third world countries is to democratize or open up their education systems.

As early as the late 1970s—or even 1980s in some countries—the postindependence euphoria was blighted by pressures from many forces: internal conflicting situations and contradictions; loss of impetus of populist nationalist ideologies; exploitative and unequal relationships between the Northern and Southern countries and undesirable side effects of reforms.

Nowadays, restraint and pragmatism have become so predominant that some education systems are at a loss as to where they should direct their efforts and priorities. The reason for this undesirable situation is that inherent in popular ideology is the view that all humankind has equal social/political rights and any differentiated approach is perceived as negative. Rulers and policy makers have transformed this into a form easily understood by people in general; therefore it has long been accepted by society.

Social forces uphold the view that the state, especially the nation-state, is the "guarantee of equity." Therefore, local school forces—under the impression that the state's decision making (as representative of public interest) is based on principles of equity—have a free hand to appropriate the education system. Consequently, there is failure to apply specific and differential rules to different sectors of the population.

The educational concept is based on a juridical model viewing the citizen as a legal subject and the nation as the outcome of rational individual agreements (Kintzler, 1988). Education is part and parcel of democracy, and citizens should be taught to express severe judgment and any shortcomings of their education system; this is fundamental to the system's vitality and survival. In fact, while education also contributes to a fertile social differentiation, under certain conditions it may give rise to contrary effects.

In this chapter an attempt has been made to show, using the example of a developing country, Algeria, how a sociocultural context that assigns only a *secondary role to women* is producing the *opposite effect in its education system*: girls *predominate* over boys in education (but for how long?). Unfortunately, even at the present time, this phenomenon is still being counteracted by society at large as when girls leave school to assume their future independence—social, private lives, and especially *gainful employment* in the world of work—these responsibilities are assumed for them.

Today inequalities of access to education, academic success, and employment opportunities are major "dramatic" factors in the lives of Algerian girls and women! The most pertinent question is, How might an educational policy, built on the principles of equity and aligned with democracy, promise and therefore create better living standards and educational management systems for all?.

Knowledge, Democratization, and Equity

Countries wishing to be part of globalization are requested to "incorporate a reference to the governing principles of equity, transparency and non-discrimination" in their legislation when they sign up with the World Trade Organization's (WTO) agreement for membership.

The Synthesis Report on "Trends and Developments in Higher Education in Europe" since the World Conference on Higher Education, 1998–2003 (WCHE) notes that a major concern for many countries is the problem of equal access to higher education and to education in general. At the root of this problem is a lack of "a meaningful increase in financial and material resources to deal with quantitative growth" (UNESCO, 2003).

On gaining independence, and especially during the 1970s, third world societies made enormous financial investments in higher education, with the intention of training managerial and administrative staff needed for future development. Emphasis was put on investment in *planning* rather than in *industry*. Technology transfer was to be the primary means whereby these countries would achieve take-off and catch up on what was seen as economic backwardness.

The economic crisis in the 1980s caused a different development model to emerge, one in which market forces—rather than public authorities—were the driving dynamics under the auspices of the International Monetary Fund (IMF) and the World Bank (WB). Populations were overwhelmed by continual "structural adjustment plans," and it came to light that economic growth and welfare programs were giving way to the struggle against poverty.

In essence, World Bank studies called for public expenditure to be restricted, *especially financial output* on higher education. The sponsors turned their gaze to, and took renewed interest in, the "social dimensions of growth" resulting from the catastrophic effects of "structural adjustment plans," that is, (a) dramatic increase of poverty; (b) catastrophic political instability, and (c) wars that frequently stemmed from such upheaval. Since then, social stability has been regarded as a "growth" factor, because it appears to result from a consensual political climate and democratic practices, built on equity. Nevertheless, education is an essential part of that equity.

Given that the financing of education is, essentially, the product of a public collective social effort, the community should in justice be able to take advantage of educational provision.

Equity in Theory and Reality

Sall and De Ketele (1997) put forward a concept of equity in education that rests on four principles:

1. Educational supply and demand
2. Educational process
3. Education of students
4. Socio-occupational lives of individuals

Nacuzon Sall (1996) defined equity as follows: "Equity is both the distribution to different social groups of opportunities for access to existing bodies, and the distribution of the costs and advantages of education and access to the same quality of education, so as to reduce the divide that exists between individuals and groups in society by improving their socio-occupational living standards and enabling them to find fulfilment in their working, social and private lives." On that basis, the strategy put forward for the practical implementation of the principle of equity should be based on

- improving conditions of study and access to learning (the social class effect);
- improving qualifications of the teaching staff (the teacher effect);
- improving management of establishments (the school effect);
- improving conditions of subsequent access to employment; and
- improving social living conditions.

Democratic principles are linked to the concept of equity, both as a value and as a pedagogical practice: the link between equity and social justice is to be found in much that has been written elsewhere.

"The essence of equity lies in going beyond proportionate treatment to put in place policies, rules, procedures and action plans that correct the gaps between members of the most advantaged groups and the less privileged. Equality calls for the equal distribution of goods in return for the same work (equal work, equal pay), i.e. proportionate treatment." "Equitable treatment is that which allows groups that in the past were affected by discrimination to accede to genuine equality." (Gaudet and Lapointe, 2002). Above all, it is without doubt, that the influence of the child's sociocultural circumstances weighs upon his/her academic success at school.

Taking account of each and everyone's needs in the process of education and learning is the fundamental principle of teaching based on equity. Cognitive theories (Vygotski, 1997) and the teaching of success and the highlighting of the multiple forms of intelligence (Gardner, 1993) have all helped to supply conceptual frameworks for approaching education in terms of equity.

Some researchers have concluded that "modern education is antidemocratic when it refuses to respect individual and collective differences." The International Development Bank (IDB) bestows its cooperation and its loans on national strategies that aim to take up the challenge of promoting equity, keeping in mind that economic growth can benefit all sectors of the population, particularly by improving educational opportunities.

One pertinent question is, Since Algeria is not a unique case, how do matters stand in the Arab states as regards the problems of equity?

Assuming that knowledge is the keystone of human development and "the tool that enables wider opportunities to be offered to human beings to improve their abilities, to conquer material poverty and to build a developed society for the incipient 21st Century," the link is affirmed between the acquisition of knowledge and society's productive capacities!

A recent study entitled "Report 2002 on Human Development in the Arab States," carried out under the joint responsibility of the United Nations Development Programme (UNDP) and the Arab Fund for Economic and Social Development (UNDP/AFESD, 2002), brings to the fore a region that has been profoundly subordinated since September 11, 2001 and the war against Iraq. These 22 Arab states were analyzed with regard to three "shortcomings" regarding females: (a) the lack of *freedom*; (b) the *subservient and inferior* status of women; and (c) the lack of education in relation to *earning power*. It is observed that the education system for acquiring knowledge, with its two major components of the "dissemination" and "production of knowledge," is worse than mediocre in the Arab states. Despite the progress achieved in access to their education systems, these countries should, in no way, conceal the serious deterioration in their quality. One must take note of the devastatingly poor performance of the communication media, taking into account the known importance they put on the production and dissemination of knowledge. In this regard fewer than 18 computers per 1,000 inhabitants against a world average of 78.3 percent; fewer than 1.6 percent of the population uses the Internet, as compared with 68 percent in the United Kingdom and 79 percent in the United States; and the number of telephone lines is less than one-fifth of the number available in the developed countries. This, however, brings us to the observation that the 1960s experience of "development in technology transfer" did not succeed in bringing about any real transformations!

The 2002 report acknowledges that there is a major human capital that could provide the impetus for a real "blossoming of knowledge." But those capabilities are tied up locally with frequent instability in scientific research development policy. The lack of continuity prevents the accumulation that constitutes the essential phase to enable knowledge to be produced. Scientific research in the Arab world "apart from producing only little knowledge . . . suffers from the weakness of fundamental research and is virtually absent in the most advanced fields such as information technology or molecular biology."

From an international study entitled the "New World Values Survey" (PNUD, 2002) on attitudes to knowledge, good governance, and gender equality, it emerges that it is the Arab states that express most strongly the view that democracy is the best form of governance, and that knowledge is a strong value to be upheld.

While gender equality in education is accepted, equality in employment is not, and this is the point where an attempt will be made, further on, to develop through the example of Algeria.

Educational Wastage

In drawing on the experience of a developing country, it is felt that the theme of "equality of opportunity for men and women in access to education" is essential as it triggers off a debate on a central feature of the system and moves it on from bland discussions of generalities, figures, and facts whose social consequences have not been fully analyzed and concrete conclusions drawn.

From the time of Algerian independence, July 1962, the school system as a centralized education system was marked by two historical and sociocultural constraints:

1. The need was to instill a sense of nationality after more than 130 years of colonization—when local cultures were curtailed or even eliminated.
2. The school system should accommodate all generations.

The school today is an institution that provides real socialization, on the basis of which individuals define themselves as educated or noneducated, qualified or nonqualified, skilled or unskilled.[1]

A massive enrollment program, of more than 7 million pupils, caused upheavals at all levels. Unfortunately, there was no accompanying investment in research as this would have allowed the way it was functioning to be better understood—with a view to producing objective remedies. The education and train-in system, now freed for two decades from the requirements to respond well and maintain social peace, was regularly faced with the pressing need to cope with successive new academic years. Increasingly managed by the rule of numbers, it seems to have fallen victim to a kind of "inevitability" that was used as the "ultimate reason for all failures to deal with the problems" that were steadily coming to surface. One of the big realities of the education system is the issue of educational wastage and exclusion, which arises from the combined effects of four factors: (a) inappropriate teaching methods; (b) lack of a regulatory framework; (c) the derailing of the vocational training system; and (d) the social images of success.

Wastage and Exclusion
Wastage and failure rates are indicators of the difficulties of raising a heterogeneous school population to a *given success rate in a given time*. Thus, in basic education we have a 13.9 percent wastage rate (15.1 percent for boys, 12.2 percent for girls) with a Basic Education Certificate (BEF)[2] success rate of 30 percent and a Baccalaureate success rate below 20 percent. In fact less than one in 15 pupils who enter the first year of schooling continue on to the final year (CNES, 1995). Teaching methods promote and build on the regurgitation of knowledge passed on by teachers, who are themselves required to regurgitate the information contained in the teachers' educational files. It was only in 1996 that the ruling to refer to the teachers' files was abolished. Naturally, the lack of provision for assessing teaching methods resulted in unawareness of teaching content and methodologies. Almost 13 years (Lakhdar, 1999, MNE) of school education, rather than the 9 normally needed, are required to take a pupil to the ninth year of basic education, or school-leaving age.

So only 8 percent of pupils in any given cohort obtain their school certificate without ever having repeated a year, and 28 percent obtain it after having repeated 1 or more years. In all, 64 percent of pupils in that cohort leave school without a certificate.

Exclusion from Regulatory Arrangements
Thus the factors that might make the system effective or play a regulatory role have not been encouraged or even introduced. The scant interest taken in preschool education—which is essential to standardizing the entry level to

the first year of basic education—has heightened inequalities and their corollary, exclusion. Similarly, the corps of teaching advisers has been abolished. This has burdened inspectors with more administrative and supervisory functions, to the detriment of their pedagogical function of monitoring, and this in turn has led to inadequacies in teacher training. The failure to assume institutional responsibility and the marginalization of special classes—catering for pupils who are failing—has led to extreme disparities in level, which make the classes unmanageable for the teachers. The canteens and boarding facilities that used to support needy families have been closed, and this constitutes an additional factor of exclusion and wastage.

The Marginalization of Vocational Education
Although in year 2002 the Ministry for Vocational Education and Employment (MFPE) (Ministère de formation et de l'emploi) was running 708 vocational training and apprenticeship centers catering for 167,980 young people, it should be pointed out that the number of rejected students and dropouts currently amounts to a critical 400,000. The present-day vocational training network—which is mostly public but includes private establishments—cannot cope with such a large number.

With the establishment of Universal Basic Education (UBE) the academic standards of pupils, entering training establishments, escalated higher and higher, but their training programs and staffing changed very little.

The narrowness of the employment market, together with the low rate of enterprise creation and the reduction in public sector employment, make the future of graduates from vocational training centers even more problematic (only 10 percent of public vocational training center graduates find employment, according to an MFPE report).

Vocational training in its current shape responds neither to the needs of the productive sector nor to the demands of those excluded from the school system—supply fails to meet demand.

The Effect of Context

The predominant social image of success—in concrete terms, university entrance—combined with a school system methodology in which selection operates through failure result in the fact that leaving school, in any shape or form, is interpreted as a failure. Vocational training has not been unaffected by the massive schooling drive, insofar as its core function of "training a skilled workforce to meet specific socioeconomic development needs" has been transformed into a "social function of providing a shock-absorber for educationally-failing pupils." In fact, "vocational training" *lost its soul* when confronted with the need to manage such an influx, and it gradually has been transformed into the "management of training places" for an ever-growing population. The result is that vocational training has become a satellite of the school system (Benghabrit-Remaoun, 1999). Socially devalued, vocational training has become identified with the, shunned, manual worker status. Additionally, the lack of specific training in crafts means that young people do not appear to have any serious training plan in mind.

Subsequent experience has strongly confirmed the links that many researchers established, as early as the 1960s, between unequal academic success and social inequalities, but the differences in the solutions found have been based on whether pedagogical or family factors are considered primarily responsible. Opening up the education system to everyone and unifying the system are no longer enough to guarantee equality of opportunity. Worse still, the school has become a place where people's disregard for the manual trades, and agricultural work in particular, is reinforced (Haddab, 1982).

The outcome is a gulf between life and school. Only just under 10 percent of young people between 16 and 18 years of age can enter vocational training establishments—the others, officially either too young to work or to be called up for national service, loiter about in the streets.[3]

The school education system, which they simply passed through, has barely had an influence on them, and was, above all, just a source of frustration. Frustration in relation to those whom the school system has selected (Benghabrit-Remaoun, 1987)[4] as having the necessary ability to continue to make their way up the ladder; in the cultural void that surrounds the school-excluded youth is nonetheless united with the "chosen ones"—unemployed, loitering together in their neighborhoods[5] they share the same dreams of social success, access to leisure, and emotional well-being.

The pupils' new aspirations are the best indicators of these changes. The rise and success of "business" strengthens and promotes "smartness" as a value: work and effort are thus devalued in their eyes.

But the key is to get your hands on money, because then *you can go anywhere and do anything.* Sad to say, the reality is that for those excluded from the system, there will be no opportunity to return to school one day. The sentence is final; there is no appeal. This sentiment of failure is reinforced by youngsters' spontaneous experience of society, and the world of work, where "business" carries more weight and the "moonlighting" economy is seen to be more lucrative. Since their educational level allows them to aim only at junior positions, progression within a workplace—if they find a job—is almost unheard of.

What remains is uncertainty and anxiety for the future, with their corollary "smartness" and the search for any way of winning a place on the social ladder, but this is also an operational characteristic of all institutions in which knowledge increasingly takes a backseat to the workings of power.

The promises of social advancement through educational qualification, which the state made during the booming days of oil wealth, have fallen by the wayside. Moreover, the reality of authority, embodied by its representatives at different levels, demonstrates clearly that an academic career is no longer the royal road to a prestigious social position. Social progress is now made through other channels, particularly political and economic ones.

Consequently, this lack of future prospects, of a positive path in life[6] that school and society have brought about in both personal and social terms has given rise to a latent detonator that is ready to explode at any time.[7]

Revolt expresses the failure of an education system that can no longer honor the promises it made. Without an escape route, without a voice in the family, at school or even in society at large, young people are left with only one space to call their own—the street, where they therefore get out of control.

But in these times of uncertainty, religion becomes once more the active ingredient of identity.

- Is fundamentalism not a form of attachment to the concept of an egalitarian technocratic school which is supposed to provide everyone with education and success?
- Has this malfunction of the education system not helped to generate that rise of religion?[8] (NAQD Review, 1993)

One thing to note is that while fundamentalism has been on the ascendant women and girls have become the preferred targets of those who think that, thanks to their educational success, women have usurped roles that were not traditionally theirs.[9]

Inequality in Access to Education and Girls' Educational Success

Although school is an ideal arena for socialization, which some researchers view as decisive in "the production line of gender and differences between the sexes," it is thought to contribute to maintaining and producing social differences between the sexes (Lescarret and de Léonardis, 1996, p. 99). At school itself, these differences between the sexes, which indeed are unequal in real society, transform the parameters of inequality into success and educational performance by drawing on the qualities of obedience, patience, and effort that are attributed solely to girls.

Unequal Educational Access and Differences in Educational Achievement between Boys and Girls

Educational enrollment in Algeria has risen steadily. The increased number of enrolled pupils, and especially the increase in the number of girls, is the most visible result (Ferroukhi, 1995)[10] of the educational and training policy pursued. Studying the trend in the rate of enrollment of 6- to 14-year olds by sex over a period of some 30 years, we observed that *girls are increasingly successful at every stage*. At the third stage of basic education (the *collège*), the proportion of girl pupils rose from 45.61 percent in 1996–1997 to 46.52 percent in 1997–1998 and to 47.20 percent in 1998–1999.

In secondary education after the certificate examination and selection for the *lycée*, the proportion of girls was higher than that of boys: 54.54 percent in 1996–1997, 53.73 percent in 1997–1998, and 54.89 percent in 1998–1999. See table 4.1.

The national enrollment rate for 6- to 15-year olds is 89.93 percent, but it is 81.61 percent for girls and 83.27 percent for boys. To take the efficiency factor, which represents the ideal number of pupil-years needed for the pupils of any given intake to complete the educational program, it should be observed that in the year 2000 the basic educational program was considered complete when the Basic Education Certificate (BEF) was obtained: the overall success rate was 34 percent; among girls it was almost 41 percent, but among boys it was only 28 percent.

Different Educational Success Rates

The gap between the sexes widens as pupils advance up the education system as seen in Table 4.2.

Table 4.1 Lycée selection: Proportion of girls to boys

All areas	Girls (%)	Boys (%)	National (%)
National	81.61	83.27	89.93
Constantine	90.20	89.67	89.93
Djelfa	59.51	64.95	62.36
Mostaganem	68.42	81.56	75.08
Thinly populated areas			
National	57.98	72.62	
Bejala	84.81	91.59	88.35
Djelfa	22.55	35.19	30.80
Mostaganem	52.96	75.26	64.27
Urban areas (chief towns)			
National	86.30	87.59	86.95
El Tarf	90.67	91.45	91.06
Tamanrasset	69.03	78.39	71.82
Mostaganem	87.15	89.05	88.11

Source: Office of National Statistics (ONS)—General Population and Housing Census (RGPH) 1998.

Table 4.2 Educational success rates (%)

Year 4	Year 5	Year 6	Year 7	Year 8	Year 9
3.36	4.28	5.91	10.74	15.85	12.61

Source: Benghabrit-Remaoun and Lakjaa (1994).

The efficiency factor is 55 percent for boys and 71 percent for girls. Therefore, girls' academic success and certificates of qualification should enable them to have access to gainful employment without any discrimination.

Unequal Access to the World of Work for Women

Despite the efforts that have been made and just pointed out as regards educational enrollment, it remains true that Algeria is the country with the lowest employment rate for women, and that even among countries acknowledged to have close sociocultural comparisons, that is, the three Maghribi states (Algeria, Morocco, and Tunisia). However, the "research development of statistical theory" (CRASC, 1995; IMED, 2000; Insaniyat Review no. 1, 1997; Insaniyat Review no. 4, 1998a; Insaniyat Review no. 6, 1998b, amongst others) testifies to the efforts being made to grasp the reality of the employment of women outside the public sector. The crisis of the welfare state, and the emergence of a new social order in which the state has shed its economic decision-making powers, have imposed a different way of governing society.

According to the international classification of countries by volume of women gainfully employed, until the early 1990s, Algeria occupied the next-to-last place. This situation cannot indeed fail to be related in part to the "turn-a-blind-eye" attitude of the national statistical establishment. Suffice it to recall the conceptual "shifts" produced by the Office of National Statistics (ONS) (Office National des Statistiques de l'Algérie), as the reality of undeclared, informal women's employment finally dawned on it: in global terms, the ONS's Questionnaire moved from the concept of "women as housewives" (General Population and Housing Census [RGPH], 1996) to the concept of the "women in part-time employment" (FPO-RGPH-1997) and then to the latest, "Home Workers," of which over 96 percent details are women (Survey Title: Labour Force and Demographic Survey, 1990—[Enquête main-d'oeuvre et démographie (MOD)]. Women 'Home Workers' recorded stand on average at 180,000 and most of them are married; as compared with women in employment, most of whom are single (Benghabrit-Remaoun and Lakjaa, 1994).

However, these same statistics point to women making an ever-increasing demand for employment during the 1990s. Now, bearing in mind that this decade was stamped by the consequences of the fall in oil prices and the introduction of structural adjustment, one cannot but ask the question, Is the increase in unemployment among women cyclical—or does it reflect a social change dynamic of which one of the most relevant indicators is employment for women?

The Cruel Reality of Employment for Women

Compared with the size of the active male population in December 1992 (74.18 percent), the active female population is notably almost marginal at 8.75 percent.

In Algeria, official paid female employment has the following characteristics: youth, the gap between the trained and the untrained, concentration in the capital city (Hakiki, 1980); and the fact that it is largely confined to unmarried women (67.7 percent) (Insaniyat Review, 1997).

In 1992, the Office of National Statistics (ONS) counted 364,338 female employees, of whom 55,883 were in the private sector (15 percent of the paid workforce).

In December 1991, 360,000 women (not counting the 158,000 home workers) declared themselves to be employed, which amounts to 8 percent of the population in employment. The majority—69.4 percent—of these women in employment were aged between 20 and 34 years. Of these, 20.2 percent were between 20 and 24 years; 31.5 percent were between 25 and 29 years; and 17.7 percent were between 30 and 34 years. Taking a different criterion, only 31 percent of them described themselves as married, with 53.7 percent being single and 14 percent widowed or divorced. So 67.7 percent of them were unmarried. In 1992, the ONS counted 364,338 female employees, of whom 55,883 were in the private sector (i.e., 15 percent of the paid workforce).

Paid employment accounts for the majority of women in employment (87 percent in 1991). In 1992 in the national public sector (industry, administrative offices, transport and communication, and commerce and service), "women accounted for just 7.62 percent of the total workforce estimated for the survey, of which 93.89 percent were permanent employees, broken down into 62.29 per cent executive staff, 25.16 per cent supervisory staff and 11.23 per cent managerial staff."

The other significant characteristic of female employment is that employed women are markedly better educated than their male counterparts. Whereas 86 percent of them had attended school, 41 percent of them had secondary education and 22 percent of them had higher education. The distribution of employed women in the labor market appears not to reflect their *educational progress*. According to the Survey in December 1991 of the Office of National Statistics (ONS), they were preponderantly concentrated in the tertiary sector: 28 percent in basic education, 24 percent office staff, 12 percent in unskilled employment (cleaners), and 7 percent in paramedical services. Women in industrial employment remain relatively scarce at 6 percent.

Differences in School-to-Work Transition by Sex

In view of rising unemployment among young people, the Ministry of Labour and Social Welfare took a number of measures aimed at allowing them to gain work experience. Three types of activity were undertaken: (a) encouraging the setting up of small businesses; (b) paid employment in local initiatives; and (c) training schemes. See table 4.3.

The results of these steps recorded by the Ministry of Labour in 1994 testify to

1. low female employment;
2. the virtual exclusion of women from some sectors; and
3. a strong presence in government departments

Although women are far less well represented in setting up cooperatives or microenterprises, they form the majority of those who benefit from training schemes.

Analysis by sector and by sex (see table 4.4) of the distribution of jobs in local initiatives shows that women are well represented in the service and

Table 4.3 Distribution by type of activity and sex

Type of activity	Total	Male	Female	%
Small businesses	4,143	3,372	771	18.6
Local paid employment	69,888	65,902	12,986	18.6
Training	2,340	745	1,595	68.1
Total	76,371	70,019	15,352	19.0

Source: Benghabrit-Remaoun and Lakjaa (1994).

Table 4.4 Distribution of local paid employment by sector and sex, 1994

Sector	Total employment (1)	Women's employment (2)	January 2, 1994 (%)
Agriculture	13,749	162	1.18
Construction	41,461	536	1.29
Industry	1,543	87	5.64
Services	41,056	5,853	14.25
Administration	17,773	7,429	41.79
Total	115,582	14,067	12.17

Source: Benghabrit-Remaoun and Lakjaa (1994).

administrative sectors. The qualifications of unemployed women offset recruitment practices that confine them to the traditional service jobs.

The State of Research

For the most part undertaken by universities, research into the social status of women is underdeveloped (CRASC, 1995; IMED, 2000; Insaniyat Review, 1997; Insaniyat Review, 1998a; Insaniyat Review, 1998b; amongst others). Characteristically, the studies that are undertaken are monodisciplinary rather than multidisciplinary and are relatively weak in field research; added to this are the difficulties of bringing together a real stockpile of knowledge. From whatever angle the question is approached, studies lay bare how slowly women are becoming recognized as individuals, instead of always having to "fit in" (taking the father's, mother's, or husband's advice), trying to win over, and avoiding direct confrontations. In connection with the topic of integrating women with development, some researchers have started by questioning the statistical tools that the state uses in counting fully employed women.

However, the main investigative tool "statistics" is not only incomplete but even discriminatory toward the same sex both as regards qualification and in the choice of sexist terminology (for example, housewives). "A woman must have a really outstanding diploma for her to have a chance of winning a 'specifically' male job" (Hakiki, 1980).

It was in fact the state, through the civil service—the public sector—that bought about the categories of employment that predominated during the 1970s and 1980s, and that were represented and targeted by statistics (Hakiki, 1980).

Today, however, a certain majority of Algeria's population is considered "marginal" because it cannot be classified in terms of statistics (the informal sector, housewives, etc.).

The rise in marriageable age has not helped women to win a bigger place in the labor market. Moreover, in recent years women's educational performance has stabilized (National Experts Commission, 1993).

In an ONS survey, analysis of employment seekers' attitudes to work—should a job become available—reveals the following. Employment seekers said that they would refuse a job on the following grounds:

- Poorly paid, 20.23 percent (women 27.16 percent)
- Distance from home, 18.17 percent (women 59.2 percent)
- Arduous or unsanitary, 31.44 percent (women 59.63 percent)
- Not the trade in which trained, 26.74 percent (women 42.36 percent)

The survey revealed that 92.7 percent of employment seekers enjoyed their families' support for their living costs, and that 33 percent were doing paid work.

This is even truer for women in view of the sociocultural context in which the keystone of parental and social behavior is "protection."

Researchers *left open to question* the dangers of seeing women's employment merely in terms of paid employment, which they describe as a narrow and limited vision of women's economic role in society. This view blames the slow growth of women's employment on "the weight of retrograde traditions and customs." A careful scrutiny of the working population and the low rate of women's employment require it to be considered in terms of work, not employment (Hakiki, 1980).

Questions designed to lay bare the mechanisms that cause women's entry into paid employment to be blocked reveal that alternative forms of employment are developing. Women are participants in development, but they occupy a singular place in it. "The fact that this is revealed neither by statistics nor by studies is due to the lack of a methodology that would permit this process to be brought to light" (Hakiki, 1997).

There are three ways in which women are remunerated. They are

1. family wages;
2. selective wages (single, widowed, or divorced women); and
3. unregulated wages.

Regardless of their marital status, one of the obstacles to their joining the ranks of wage earners is the "incomplete" nature of the status of individuals in society. For the school education system is partly responsible in confining girls to traditional female roles despite the innovations and transformations that have begun.

Various studies conducted in different sectors have highlighted the following:

1. The attitude to work of the women questioned depends on a combination of several factors: age, marital status, educational standard, and social background. Overall, young single women, regardless of their social background, are looking for employment outside the home. The quest for freedom from family constraints underlies this attitude. More mature women whether married, widowed, or divorced seek to combine work with running their households. Older women with no education, from a humble social background, and with difficult living conditions (such as cramped accommodation) work at home in the food industry (rolling couscous, preparing pasta, and so undertaking the primary processing of the product). In effect it is the 18- to 28-year-olds who are famished for employment.
2. Training may precede, accompany, or follow a working activity. Whether training comes as a private or public program, women make a

financial contribution to the acquisition of their new practical knowledge.
3. Attitudes to state institutions differ according to experience, ranging from an attitude of expecting nothing (which may nonetheless be influenced by training) to the expectation of total care. Among the group of women under survey, a keen awareness of the economic crisis found concrete expression in the compulsion to create their own work, stemming from the need for supplementary income or even the need to protect their own future, since marriage—which has become more or less uncertain—no longer gives them this protection.
4. Despite difficult living conditions, women are not prepared to do "anything and everything," and the family, albeit shaken, continues to play a protective role. The central role of the woman in the family restores full meaning to the links that must be established between work and household chores.
5. The family's presence is only justified thanks to constant negotiation. Women "always tolerate and put up with things" (take the father's or mother's advice), attempt to persuade, and never confront opposition directly.
6. Since identification of the activities in which women engage has brought out the different varieties of their situations, action to promote and support them must be equally diversified.
7. The fact is that the organization of work among "Home Workers" put to challenge the concept of "Home Working." For these women, through their work, more often than not have as many commitments outside the home as within it.

Conclusions

The school and training systems are real catalysts of social change. While the university through its socializing function helps to lay foundations for the recognition of the individual via the emergence of new aspirations, it remains equally true that it is an extension of socialization in the family. Studying the future plans of male and female students as part of a project—at the University of Oran—we noted that for female students those plans were structured primarily around marriage plans (CRASC and Benghabrit-Remaoun, 2003).

However, with the progress of occupational and academic plans it has become apparent that—under the combined impact of the skills and experience acquired through the education system and inherent social difficulties—a latent challenge is growing to those *preestablished orders that put women at a disadvantage* in the Arab world.

This challenge is acquired subjectively and is reflected in part in a demand for changes in the legal status of women in society, which in Algeria is governed and defined by the Family Code of 1984 that is still in force.

Notes

1. It is not surprising that, in research into the causes of the wave of violence that our country experienced from 1992 to 2000, the school system was portrayed as a sacrificial lamb. Nevertheless, it has its responsibility for implementing everything that may promote social cohesion, while making a special effort in a kind of *ijtihâd* (independent theoretical inquiry).
2. *Brevet d'enseignement fundamental* (BEF) or Basic Education Certificate that testifies to nine years of school education. The baccalaureate can be taken after 12 years of study, 9 in basic education followed by 3 years in secondary education.
3. The law of apprenticeship, which encourages small enterprises to take young people on as apprentices without commitment to subsequent employment, has been very little taken up.
4. The changing makeup of society has brought changes in the educational strategies of the social groups. At first and until 1980, the Arabized sections in the secondary schools were rejected in favor of the bilingual sections: when Arabization became widespread, the criteria of excellence and choice became the mathematics and science streams.
5. The most widespread local expression is *hittiste* (a colloquial term for the unemployed).
6. Difficulties in finding a job because of the "crisis," impossibility of having your own accommodation, which makes marriage unthinkable, the lack of local cultural life allowing people to enjoy their youth, and alongside them the ever-present model of the easy life and success represented by the rise of the speculative bourgeoisie. That is why young people refuse to accept that the climate of austerity imposed by the state is legitimate, since it is not shared by all social strata.
7. The riots of October 1988: there had been several warning shots. Before the October 1988 events there was Tizi Ouzou in 1980 and Oran in 1982, a movement of secondary school pupils that was very quickly supported by all the potential forces of the excluded and the underclass, Constantine and Setif in 1986. These movements were violently put down.
8. In this regard, reference can be made to the partial results of an inquiry that we have carried out—and published—among final-year secondary school pupils.
9. At book fairs such as those of 1985–1986, massive imports of Islamic books were to contribute to structuring the editorial arena with a social impact that was felt in the 1990s.
10. Opportunities for access remain unequal in geographical terms, with urban areas better placed than rural areas, and in socioprofessional terms.

References and Works Consulted

Benghabrit-Remaoun, N. (1987) Contribution à une étude des aspirations des lycées du technique. In round table *Ecole et idéologie*. Centre de recherche en anthropologie sociale et culturelle (CRASC). Oran: University of Oran Es-Senia.

—— (1999) Politique de formation et mutation: les paradoxes d'une réalité. In a collective work directed by A. Djeflat (ed.) *Algérie de novembre à l'ajustement structurel*. Dakar: Codesria, Karthala.

Benghabrit-Remaoun, N., and A. Lakjaa (1994) *Etude sur l'emploi féminine*. Ministry of Labour (MOL).

CNES (1995) *Report on the Education System*. Algeria: Conseil National Economique et Sociale de l'Algérie/National Economic and Social Council.

CRASC (1995) *Femmes et développement*. Oran: Editions CRASC.
CRASC, and Benghabrit-Remaoun, N. (co-ord.) (2003) Les étudiants de 1ère année: savoir, culture et environnement. Case studies, University of Oran Es-Senia.
Ferroukhi, D. (1995) L'état de l'éducation en Algérie. *Statistical Collection* (56), Algiers.
Gardner, H. (1993) *Multiple Intelligences: The Theory in Practice*. New York: Basic Books. Editions Retz (French translation, 1996).
Gaudet, J., and J-C Lapointe (2002) L'équité en éducation et en pédagogie actualisante. *Education et francophonie*, August 2002.
Haddab, M. (1982) *Les jeunes ruraux et l'école*. Editions Research Centre of Applied Economics for Development (CREAD), Algiers.
Hakiki, F. (1980) Le travail féminin: Emploi salarié et travail domestique in Actes des Journées d'étude et de réflexion sur les Femmes algériennes. Cahier du CDSH, no. 3, Oran, May 1980.
——— (1997) Les statistiques de l'emploi et de la formation en Algérie. Une approche en terme de convention. In M. Bougroum and F. Werquin (eds.) *Education et emploi dans les pays du Maghreb, ajustement structurel, secteur informel et croissance*. Editions Marseille, CEREQ, doc no. 125.
Institut méditerranéen (IMED) (2000) *Les Algériennes, citoyennes en devenir*. Algeria: Editions CMM.
Insaniyat Review (1997) Le travail informel: Figure sociale à géométrie variable (le travail à domicile). In *Revue Algérienne d'anthropologie et des sciences sociales no. 1*, by A. Lakjaa. Oran: Editions CRASC.
——— (1998a) Familles d'hier et d'aujourd'hui. In *Revue Algérienne d'anthropologie et des sciences sociales no. 4*. Oran: Editions CRASC.
——— (1998b) L'école, approches plurielles. In *Revue Algérienne d'anthropologie et des sciences sociales no. 6*. Oran: Editions CRASC.
Kintzler, C. (1988) L'enfer est pavé de bonnes intentions: esquisse d'un concept de l'école fondée sur une philosophie du droit. In *L'école de la démocratie*. Editions compiled by Renou, X. Fondation Diderot.
Lakhdar, B. (1999) Basic Education: Balance Sheet for the year 2000. Ministry of National Education, Baghdad, October 1999.
Lescarret, O., and M. de Léonardis (1996) Séparation des sexes et compétences. Paris: l'Harmattan.
NAQD Review (1993) Nouria Benghabrit-Remaoun, Jeunes en situation scolaire, représentation et pratiques. *Revue d'étude et de critique social, no. 5*. Contribution to dossier *Culture et système educatif*. Algiers.
National Experts Commission (1993) *Présidence de la République*. Algérie, 2005.
OMS (Office National des Statistigues de l'Algérie), General Population and Housing Census (RGPH), 1990–98.
PNUD (2002) Rapport sur "Le développement humain dans le monde arabe." New World Values Survey.
Sall, H.N. (1996) Efficacité et équité de l'enseignement supérieur. State Doctorate at the University of Dakar.
Sall, H.N., and J.-M. De Ketele (1997) L'évaluation du rendement des systèmes éducatifs: Approches conceptuelle et problématique. *Mesure et évaluation en éducation*, 19, no. 3.
UNDP/AFESD (2002) *Arab Human Development Report: Creating Opportunities for Future Generations*. New York, p. 3.
UNESCO (2003) *Synthesis Report on "Trends and developments in higher education in Europe." Meeting of "Higher Education Partners."* UNESCO, Paris, June 23–25, 2003.
Vygotski, L. (1997) "Thought and language (1934)." French translation *Pensée et langage*, Editions *La Dispute*. Paris.

Chapter Five
Knowledge, Culture, and Politics: The Status of Women in the Arab World

Fahima Charaffedine

With the publication of the first *Arab Human Development Report* (UNDP, 2000), in the year 2000, the debate on the constraints and obstacles of human development in the Arab countries reached its highest point. In addition to the obstacles formed within the regional and global realm, in the shadow of the imbalance of power and the clear noncompetitive status of the Arab world vis-à-vis the rest of the world, this first United Nations *Arab Human Development Report* has, yet again, put the spotlight on the inner constraints that penetrate Arab societies and hinder the necessary jumpstart of development. In fact, the lack of knowledge (UNDP, 2000), the lack of freedom, and the lack of women's empowerment are still the basic impediments of any resurgence.

The claim to the importance of this report does not lie in its qualification of the right facts—and the facts are right—but in its urgent invitation for the reconsideration of the real problems that, since the early nineteenth century, formed the focal point of the interest and preoccupation of the Renaissance thinkers. We only need to consider the contents page of Rifa'ha Rafe'e al Tahtawi's book *Takhlis el Ibriz fi Takhlis Bariz* (1987) to see how these Renaissance thinkers faced the problems with no fear or restraint. The call for education and acquired knowledge (education of girls and boys), the call for freedom and the establishment of reputable governance as opposed to despotism, the call to educate women and elevate them to the level of men (made by Kassem Amin toward the end of the nineteenth century) (Amin, 1899), have been at the core of demands centering on women until today.

The realization of these demands has seen some progress, especially in the independent nations created after World War II. However, the result did not match expectations—as the sociological and economic studies and international and regional reports confirm: "[a]spirations to freedom and democracy are still an out-of-reach wish" (UNDP, 2000) in the Arab countries and, despite the progress in educating women (UNDP, 1995) during the past 50 years, more than half of Arab women are still illiterate, and "illiteracy within Arab adults reaches 65 million, two thirds of whom are women" (UNDP, 2000, p. 25).

More then 100 years ago, the Arab social thinkers and social reformers struggled against the inherited and prevailing social traditions. Kassem Amin (1899, p. 9), author of the book *The Liberation of Women*, showed this "through historic experience, the coupling of the degradation of women and the degradation of nations and their barbarity, and the coupling of the elation of women and the elation of nations and their civility."[1]

After more than a hundred years, where is the Arab woman today? Moreover, what is her position in the Arab society?

Current Status of Arab Women

It may appear strange to talk about Arab women in the shadow of 22 nations—with the diversity of economic structures and levels of growth, differences in economic directions, disparity in wealth, and the different forms of government and authority. However, looking at the status of women and what this status reflects about a cultural and religious domination—to name but a few factors—makes the possibility of talking about Arab women in general more feasible.

Statistical studies confirm this point of view; in fact, we do not find a qualitative difference in the status of women and their positions between Lebanon, which is considered the Arab country most influenced by Western culture and its forms, and Jordan, which started its modern journey just recently. Women in Lebanon are given equal opportunities for education. The enrollment rate of gender-/social-type schooling in 1998 was 0.99 percent, and the number of graduating girls is higher than graduating boys. Yet, in Lebanon, women's participation in parliament was only 3.2 percent. No woman has been named minister or undersecretary; thus, women's participation in the executive levels of government is nil. Women in Jordan surpassed women in Lebanon in their political participation, because of the differences in the political system; the king of Jordan appoints women in Parliament, and as ministers to participate in politics; the participation of women in politics in Jordan is at the king's discretion.

Thus, the current status of the Arab woman is still problematic, and the awareness of such status is still embryonic. Change has just started, as some progress is seen in the status of women "in the arena of education and benefits of health care and employment" (United Nations, 1998). Yet this achievement is not enough, and the position of women and their participation is still below the expected level, especially concerning women's participation in public life and in decision making.

All reports and studies confirm that the hurdles against the progress of women are caused by "the traditions, value systems, customs and religious practices [that] form the only and most important cultural frame which limits the participation of women and their engagement in the process of progress" (United Nations, 1998). The UNDP *Arab Human Development Report 2000* states that "the Arab countries suffer from a clear lack of focus on women empowerment, and the Arab world ranks the region [last but one on] the scale of Gender Empowerment Measure (GEM), followed only by Sub-Saharan Africa."

The indicators given in figure 5.1 indicate the limits of the position of Arab women, especially in modern writings that link the position of women to the role of women in decision making and decision implementation.

In brief, Arab women are still in a lower position vis-à-vis men, and this despite the equality set forth by the constitutions (Nassar, 1985) of different Arab countries where equality of responsibilities and rights is confirmed

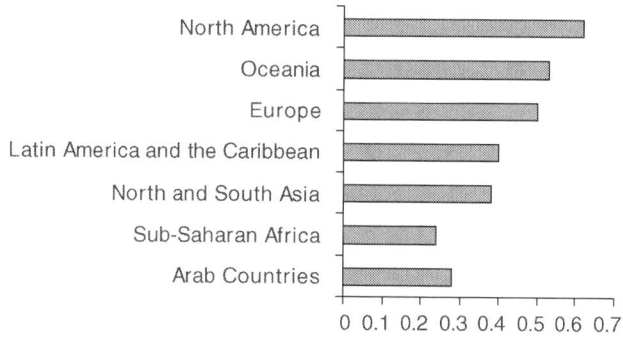

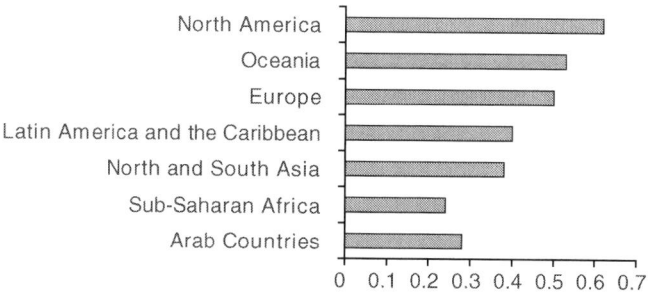

Figure 5.1 Average GEM (gender empowerment measure) values: World regions, 1995
Source: Human Development Report (2002).

between all citizens. These constitutions themselves have kept the personal status laws inherited from the Ottoman Empire: personal status laws were still regulated by religion and sects, with all the inherited traditions and values, which were transformed to beliefs directing not only the relationships of men and women, but also social and political relations.

It seems that the lack of female empowerment goes beyond women's rights in equal opportunities and equality, to reach the issue of sustainable growth in the Arab world. We claim that the causal relationship between gender and social type and growth on the one hand, and the causal relationship between these issues and freedom and knowledge on the other, accounts for the "logic of crisis" that prevails in Arab societies. In fact, the lack of freedom, and the absence of a bill of rights that is conditioned by reputable governance, prevent the possibility of an adequate knowledge structure for building the required human capacities, among them women's capacities and women's empowerment. It is worth noting that the conditions that appear to be objective are deeply influenced by the subjective conditions regulated by the values and rules of social behavior in the Arab world.

It is important to admit that women are facing constraints and obstacles that hinder their participation in public life. These obstacles are apparent, but in our view, they are rooted directly in the historic and cultural details that contributed to the building of the Arab social vision of women and their roles. These obstacles are thus an essential factor in the unification of the vision of Arab societies and the social position of women.

Researchers agree that these "objective" obstacles are in fact built upon subjective social obstacles structured on the core of the Arab cultural system. We could categorize these obstacles on three different levels, in tune with the requirements of sustainable growth:

1. The first obstacle is illiteracy: in fact, female illiteracy is still very high, reaching 15 percent (females above the age of 15) in some countries, and upto 76 percent in others (United Nations, 2000). Average female illiteracy is always double that of male illiteracy. This average confirms that illiteracy portrays the level of growth in Arab countries, and also points clearly to the discrimination against women in the essential services (education). This discrimination, still seriously present in Arab societies (Charaffedine, 2002), means that the mobilization in favor of the importance of women's education, and their position and participation in public life, is always very weak in these societies. See figure 5.2.
2. The second obstacle is fertility. In Yemen, for example, the fertility rate is the highest in the world (UNDP, 2000). The fertility rate is linked to the whole set of social and cultural practices, such as early marriage, gender discrimination, illiteracy, and the restricted awareness of the importance of family planning. In these circumstances, it is impossible for women to govern their own bodies and to decide for themselves on matters of pregnancy and childbearing. Traditions, the patriarchal value system, and religious beliefs play a major role in the establishment of inequality between men and women.
3. The third obstacle is the rate of women's economic participation: the rate of economic participation of Arab women is considered to be the lowest in the world (ESCWA/Cawthar, 1998). In fact, whilst the overall Arab countries' average is 17 percent, the economic participation of Arab women varies between 7 percent in the Arab countries such as Qatar and Saudi Arabia and 28 percent in others (e.g., Lebanon). Moreover, the structural breakdown by type of work (see table 5.1) reflects the social perceptions of the role and position of women and determines the areas where women work.

The three obstacles—illiteracy, fertility, and economic participation—interact organically with culture and politics. Researchers and social analysts agree that these obstacles delineate the general direction of the position of

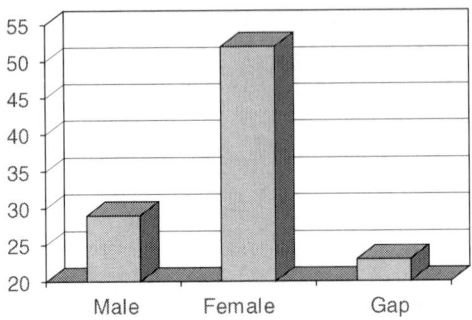

Figure 5.2 Regional illiteracy rates by gender and gap (for early 1990)
Source: *Women and Men in the Arab Region: A Statistical Portrait* (2000).

Table 5.1 Structure of active women economically in professional, technical, administrative, and managerial posts in Arab countries (1970–1994)

Country	Year	Professional, technical, administrative, and managerial (%)		Professional, technical administrative, and managerial Number	
		Total	Women	Total	Women
Whole region	1970	5.8	10.6	1,429,487	269,276
	1980	9.2	13.7	3,437,403	764,983
	1994	10.5	12.9	5,270,898	1,301,183

Source: ESCWA (2000).

women in Arab societies, and contribute to the affirmation of the roles of men and women.

With such obstacles, how does the status of women manifest itself? Moreover, what are the possibilities and prospects for change?

The Status of Arab Women

In modern writings about women, the concept of women's status is linked with the capability of women to participate and intervene in the process of decision making; this usually symbolizes the changes in women's positions, capacities, and capabilities.

If the concept of women's status takes for granted certain knowledge and cultural conditions; it also assumes a legislative and political environment that allows for a legal status based on equal opportunity and justice. Would Arab societies meet these criteria?

To say that Arab societies are passing through one of their toughest phases is not new. In fact, besides the lack of aspects that were highlighted by the *Arab Human Development Report*, Arab societies also suffer from another more strenuous problem: the persistent reluctance to assume the "moral" responsibility for the life conditions of these societies.

To continue the old analysis (where colonialism is considered solely responsible for present life conditions) will not solve the problem; true, colonialism may be responsible for the "underdevelopment" that still engulfs the Arab world. Moreover, to hide behind the alleged specificities of Arab societies in order to postpone the discussion of urgent issues, such as human rights or women's claims, will not solve the problem either.

The knowledge situation in Arab countries has reached a crucial point where surgical interventions are needed in the three areas concerned with improving knowledge, producing knowledge, and benefiting from knowledge (UNDP Regional Bureau, 2003). Whilst knowledge cannot be created out of nothing, the capability of building knowledge is linked directly to the social structures generating that knowledge. Talking about social structures takes us back to both culture and politics. What can be said about politics in societies still living in an age of despotism? The lack of freedom highlighted in the report not only fully describes the relationship of the governors and the governed but also describes the relationship between politics and knowledge—especially in the arenas of creation and invention, and of marginalization and exclusion. As for the legislative structure in Arab societies, a structure paralyzed by law-inhibiting procedures, it does not and will not provide any legal conditions to protect different individuals and groups.

Although Arab constitutions already grant individuals and groups equal rights, the actual practice of the laws and regulations in most Arab countries frustrate the application of those constitutions. The Arab position vis-à-vis the document for the "elimination of all forms of discrimination against women" is a proof of this. In fact, not all Arab countries ratified this document; the 13 countries that ratified the document did so with many reservations on numerous articles related, in particular, to the personal status, the laws of nationality, and some articles related to the penal code—all articles related to what we call the independent entity of women. With such reservations, the countries have blocked the concept of equality, which is the core of the document. We believe that political authorities in Arab societies do not respect the constitutions and do not insist or struggle for the respect of the constitutions, as is apparent in the position of various women's movements.

The constitutions are only drafts with no practical purpose, and the laws are in no better situation. The laws of emergency, and the temporary laws prepared for times of war outlive them in times of peace. These laws survive with the survival of the authorities, and may continue with subsequent authorities, despite the changes in political regimes and its representative (e.g., Egypt).

The cultural discourse adds to the existing ambiguity around the major issues, which remained absent from social thinking till the 1990s

(Bechara, 1992)—such as human rights, civil rights, and women's rights. The discourse on cultural specificities is also ambiguous, for it always pushes the arguments toward imagining these peculiarities as belonging to a world apart from our own (the Arabs), yet the specificities are not particular to the Arab world. In fact, the mediocre situation of women is not an Arab or Muslim specificity, but it is part of the history of women throughout the world. Indeed, unfulfilled current needs of Arab women are similar to those of Western women at the beginning of the nineteenth century.

The simplification of the problematic related to culture, and its treatment as a contradictory positive/negative specificity, does not allow a neutral understanding of this specificity. In fact, specificity and universality are not absolute concepts. In this respect, it is imperative to debate the concept of specificity in order to separate issues such as human rights or women's rights from the closed circle of specificity that is forbidding any kind of interpretation or analysis.

It is time to take a critical stand vis-à-vis this alleged specificity, because it is not true that the position of women in Arab countries is reinforced by women being wives and mothers. No matter how much we exaggerate the glorification of motherhood, motherhood remains an elective entity, an option, of women, and not their whole or entire entity.

As for the position of wives, it is enough to look and listen to appreciate the depth of women's pains and grievances (Charaffedine, 2002) in the shadow of the codes of personal status that regulate the relation between men and women, and issues of marriage and divorce. It is enough to look and listen to define who does what and who has what, to understand the position of women in the Arab World.

What is Missing, and Trends of Change

We may not need many details to explain the subordinate position of Arab women today. The lack of women empowerment is made clear when considering the relationship of women to authority/power. In fact, empowerment indicates the relationship between women's capabilities and their position. According to UNDP's *Human Development Report* (1995), the measurement of gender empowerment reflects the capability of women in participating in political decision making, as well as in women's job opportunities and earning power. Though the UNDP measurement of empowerment does not cater for other dimensions of women's empowerment—such as empowerment in the family or in local and regional societies—it could be considered as a measurement of the organic relationship that is built between knowledge, culture, and politics on the one hand, and the position of women on the other.

Statistics provided by national reports pinpoints to the weakness of the relationship of women to authority/power. Despite the equality of political rights imposed by constitutions between men and women, women's participation in all political structures in the Arab world is still very low: the rate of women in Arab parliaments is the second lowest in the world (after sub-Saharan Africa).

The participation rate in executive governmental bodies is no better. The participation of women is considered as a way to improve the image of political regimes, especially under the international campaign to improve the position and status of women.

Both in Egypt and the Syrian Arab Republic, two ministers are women; women occupy ministerial positions in Morocco, Oman, and Qatar. However, these women are ministers either of culture or education, tourism or social affairs; the limitation of women to such ministries reflects the perception of Arab societies about women's capabilities and their roles in public life.

It is also unacceptable for women's participation to be so low in countries where women's capabilities are equal to those of men. For example, in Lebanon, the number of female college students is higher than the number of male college students, yet no woman has ever occupied a ministerial position since the creation of the modern state of Lebanon in 1943. Moreover, women's participation in Parliament is only 3.2 percent (Charaffedine, 2002). Growth indicators confirm that the acquisition of knowledge is a major factor of change in the human development process; however, the acquisition of knowledge fails in the case of women, given the intervention of the dominant culture, which distributes the roles between both genders.

The quasi-majority of researchers agree that the reigning system of patriarchal values has the greatest responsibility for the low position of Arab women, and that power distribution in the family is responsible, to a large degree, for the imbalances in the power distribution in the political domain. Accordingly, the process of role distribution in Arab societies appears not only as a result of objective conditions that provide for women's access to education, acquisition of knowledge, job opportunities, and health care; these objective conditions, although necessary, are not sufficient to explain the status of women. The distribution of roles between men and women is based on perceptions generated by a society affected by traditions and inherited value systems. These perceptions delineate the limits of expectations of the process of "social adaptation," which starts at the family level, and the social foundations and institutions (schools, media). These expectations are the result of an education system that regulates the rules of social behavior in tune with the value system and religious codes, and form the "sacred" status of Arab women. These expectations are also the result of the legal system that defines the limits of responsibility, both in obligations and rights.

In Arab societies, where the role of women is identified only in the reproductive role, and where women are denied their productive role and their participation in political organizations, the culture of gender discrimination is apparent. This culture of discrimination is based on the educational religious traditional system, responsible for the imbalances in society.

The position of women in the Arab world is not only a result of objective conditions, but is also a hostage to these conditions. These conditions are actually a group of necessary factors for growth (education, jobs, health care) and a group of laws and legislations that impact the lives of women. The subjective conditions are built upon traditions and value systems, forming what we call the "educational system." These conditions are still under the control of religious values and infiltrate the social foundations and institutions (family, school, media) to contribute deeply in forming the image of women and men, and their roles and the positions that they occupy.

Knowledge, Culture, and Politics: Role in Changing the Status of Women in the Arab World

Statistics provided by the reports on human development clarify social trends in all regions of the world. The quantitative figures allow us to ask questions beyond the description of facts, to grasp the real reasons for the low status and position of Arab women, and what may also be the real reasons for the underdevelopment of Arab societies.

Hisham Sharabi (1985) confirms that the patriarchal structure in Arab societies, which allowed the male domination of women, generates a relationship of dependence, not only between women and men but also between the governor and the governed. Thus, the patriarchal structure is responsible for the absence of democracy, and the reign of despotism in Arab societies.

The structure of dependence that Sharabi has talked about is present in the relationship of women to power, starting from the family power of decision making to the participation in this decision making, to the highest levels of political power. Sharabi (1992) points that the concept of participation requires a strict adherence to the democratic family traditions, an adherence not present in the Arab family.

In a report to the Centre for Training and Research for the Arab Women, "Cawthar," data analysis shows that participation requires the continuity of the operations of social preparation, starting from the family where the factors of despotism are built in through the authority of the father and his relation to the family members, and "through educational institutions and the media, where political measures and the concept of a just government and the rights of individuals take a practical shape" (Centre for Training and Research for Arab Women, 2002).

Moreover, the patriarchal structure reflected in the political structure has a negative effect on the political culture and creates what we call the "patriarchal politics," which leads to despotism, or transforms the political rights into mere forms, with no real application of these rights.

In this spirit, we may understand how Lebanese women won the right to vote in 1953 but could not participate in Parliament except in very low numbers (3.2 percent). We may also understand the weak position of Arab women in everything related to decision making and its implementation.

The reasons that inhibit the position of women are different from one country to another; they are related to objective conditions such as education and knowledge acquisition, and jobs and wealth accumulation. These conditions become less efficient when confronted with cultural and historic conditions under which the perceptions of Arab citizens are built.

The best proof is to consider the status of Arab women today, which is very similar throughout the Arab countries, despite the differences of objective conditions in Arab nations.

We believe that improving the position of the Arab woman requires objective conditions related to knowledge, jobs, and empowerment, as well as other conditions, such as the building of democratic political life, and a democratic social culture open to modern knowledge and culture.

We will not propose alternatives, as it is impossible to change culture without changing its dynamic forming factors. These dynamic factors, as Ibn Khaldoun said, are people's customs and practices in daily life. These dynamic factors are the opportunity for change, an opportunity produced by the possibilities of development (in the modern meaning of the term): the development of human, economic, cultural, and social resources. These allow different images and perceptions of women, and thus different expectations and different roles for women, governed by the concept of equality, based on the capacity of Arab women and their real capabilities in the participation in governing nations and cultures, as "equality is not a technical objective, but rather a full political engagement" (Smith, 1995).

Note

1. Kassem Amin was one of the important Arab personalities in the Arab reform stage in the nineteenth century. He was interested in women's issues, especially education and work.

References and Works Consulted

Amin, K. (1899) *The Liberation of Women*. Cairo: Al Tarakki Press.
Bechara, A. (1992) *The Civil Arab Society: A Critical Study*. Beirut: Centre for Studies of the Arab Unity.
Centre for Training and Research for Arab Women (2002) "Cawthar." In *Social Gender and Globalization: The Economic Participation of Arab Women*. Tunis.
Charaffedine, F. (2002) "One Origin and Many Images: The Culture of Violence against the Women in Lebanon." In *Women Revelations*. Beirut: Al Farabi.
Economic and Social Commission for Western Asia, United Nations (2000) *Women and Men in the Arab Region: A Statistical Portrait*. Beirut.
ESCWA/Cawthar (1998) *Statistics and Indicators of Arab Women*. Beirut.

Human Development Report (2002) New York: United Nations.

Nassar, N. (1985) Women's Position in Arab Constitutions. *Al Wahda* magazine, no. 9, June, Rabat, Morocco.

Rifa'ha Rafe'e al Tahtawi (1987) *Takhlis el Ibriz fi Takhlis Bariz*, introduction by Mohama Amara. Beirut: University Institute for Studies and Publications.

Sharabi, H. (1992) *Civilized Critique to the Arab Society*. Beirut: Center for Studies on Arab Unity (in Arabic).

—— (1985) *The Patriarchal Society and the Issue of the Different Arab Societies*. Beirut: Center for Studies on Arab Unity (in Arabic).

Smith, J. G. (1995) "Introduction." In *Human Development Report*. New York: United Nations.

UNDP (2000) *Arab Human Development Report 2000*. New York: Regional Office for Arab Countries.

—— (1995) *Human Development Report*. New York: United Nations.

—— (2003) *Report on Arab Human Development: Towards Building a Society of Knowledge*. New York: Regional Office for Arab Countries.

United Nations (1998) *Survey of Economic and Social Developments in the ESCWA Region*. Beirut.

Women and Men in the Arab Region: A Statistical Portrait (2000) Beirut: Economic and Social Commission for West Asia, ESCWA, United Nations.

Chapter Six
Knowledge, Theory, and Tension Between Local and Universal Knowledge

Roberto Fernández Retamar

In Memoriam Edward Said

For the sake of understanding, I shall start by accepting, contrary to what experts of the cosmos have tirelessly advanced, that what happens on our tiny planet—a Blue Planet, as Paul Eluard and Yuri Gagarin discovered—is capable of usurping the disproportionate description "universal." Furthermore, although it might seem to be quite different, "local" means here that which concerns the third world, underdeveloped countries (sanguinely termed "developing," which is seldom the case), and "universal," essentially, not even all the countries on earth, but rather the "underdeveloping" (i.e., causing underdevelopment) ones. For about 40 years I have been proposing the latter term—*subdesarrollante*—in vain, thus far, although I believe that it does much to clarify things. In the pages that follow I shall be looking at this basic asymmetry, which came about in 1492 and gave rise to colonialism, racism, and modernity, and which, to this day, has split the world in two (although not always with the same components).

Franco (1988) wrote, "*British intellectuals: Latin American revolutionaries* was the wording of an advertisement I once saw in the *New Statesman* in England. It summed up very nicely the separation of intellectual and manual labour, along the axis of metropolis and periphery, as well as suggesting the flow of revolutionary action into areas where people know no better than to fight. The conclusion is that the Third World is not much of a place for theory: and if it has to be fitted into theory at all, it can be accounted as exceptional or regional."

Further on, Franco (1988) added,

> Metropolitan discourses on the Third World have generally adopted one of three devices: (1) *exclusion*—the Third World is irrelevant to theory; (2) *discrimination*—the Third World is irrational and thus its knowledge is subordinate to the rational knowledge produced by the metropolis; and (3) *recognition*—the Third World is only seen as the place of the instinctual.

I shall give a final quotation from this author on the matter:

> Since I refer mainly to Latin America [which is what I shall also do], it is necessary to emphasize the crucial and constitutive activity of the *literary* intelligentsia which is empowered by writing. Because it was blocked from making contributions to the development of scientific thought, the intelligentsia was forced into the one area that did not require professional training and the institutionalization of knowledge—that is, into literature. It is here, therefore, that the confrontation between metropolitan discourse and the utopian project of an autonomous society takes place. (Franco, 1988)

The latter observation is distinctly reminiscent of the words read out in Stockholm by Gabriel García Márquez when he was awarded the 1982 Nobel Prize in Literature: a date that must be remembered, as we were still in the cold war between the United States of America, today more powerful and aggressive than ever before, and the now dissolved Soviet Union. In addition, as a simple expression of the recognition that the region's literature started to receive from the 1960s, when revolution was giving off sparks in "our America," attracting the world's attention, suffice it to recall that between then and now that same Nobel Prize in Literature has also been awarded to Miguel Ángel Asturias, Pablo Neruda, Octavio Paz, and Derek Walcott—and it appears that reasons unrelated to literature but political, different but equally unacceptable, have denied it to Jorge Luis Borges and Alejo Carpentier. The words of Gabriel García Márquez (1982) to which I refer are the following:

> Why is the originality so readily granted us in literature so mistrustfully denied us in our difficult attempts at social change? Why think that the social justice sought by progressive Europeans for their own countries cannot also be a goal for Latin America, with different methods for dissimilar conditions? No: the immeasurable violence and pain of our history are the result of age-old inequities and untold bitterness, and not a conspiracy plotted three thousand leagues from our home. But many European leaders and thinkers have thought so, with the childishness of old-timers who have forgotten the fruitful excess of their youth as if it were impossible to find another destiny than to live at the mercy of the two great masters of the world. (Márquez, 1982)

What Márquez (1982) called "our difficult attempts at social change" refer to what for Jean Franco is "the utopian project of an autonomous society," which she believes is in confrontation, visible in literature, with the "metropolitan discourse." However, as I will have occasion to reiterate, "our America" (an expression of Martí (1891) that I prefer to "Latin America and the Caribbean" although I do not reject that syntagma) has not only produced and continues to produce music, dance, arts, and much more, but also projects for social change, utopian and otherwise, and other forms of knowledge and theories. It is very difficult for the lingering metropolitan mentality to accept this; and, from the other side, the colonized mentality can only go along with the mimesis it is offered or even seeks. Refusing the first and taking on the duty and the right to create "with different methods for dissimilar conditions" led me to the temptation to entitle this paper "Ariel's Alternatives," on the grounds that Shakespeare's character can be the metaphor of the intellectual, as more than one writer has suggested. In the end I set aside the temptation, for which Oscar Wilde would never have forgiven me, so as not to skew the plan for these talks, but I must confess that I still fancy the title of dramatic origin.

To continue clarifying the terms I will use (we have experienced, and are not yet free of a period of appalling semantic mystification), I must be clear that, without wishing to state the obvious, what I mean by the "West" is the capitalist, developed world: or the world that is "underdeveloping" (i.e., causing underdevelopment) if I want to remain true to myself. This world was born in a few regions of Europe and thanks to the United Kingdom—a capitalist country par excellence for centuries—took hold in some of its former colonies, not "Western" in the European sense, such as the United States and Canada, Australia and New Zealand: people "transplanted" to use the terminology of Ribeiro, where the original inhabitants were eliminated or marginalized. It is well known that one of these former colonies, the United States, is the new leader of the West and aspires to total hegemony having made even its former metropolis a vassal. The original case (and not geographically "Western" in relation to Europe) of Japan, the only genuinely capitalist country not peopled by Europeans, warrants separate treatment and differs in quite a few cultural respects from the others of its kind. The central core of these nations, the "Big Brothers" of the moment, are grouped in the G7 (now G8). Many of the remaining countries cannot be called Western but in any case Westernized, and they provide the masters of the world with their external proletariat to use the expression coined by Arnold Toynbee.

As it seems to me to be enlightening, and because it was published in a magazine not to be suspected of radicalism, namely *Time Magazine*, I have cited on other occasions, and will do so again, the article by Elson (1992) "The Millennium of Discovery. How Europe emerged from the Dark Ages and developed a civilization that came to dominate the entire world." These lines are taken from it:

> The triumph of the West was in many ways a bloody shame—a story of atrocity and rapine, of arrogance, greed and ecological despoliation, of hubristic contempt for other cultures and intolerance of non-Christian faiths. (Elson, 1992)

Only one point requires changing in this true and brief sentence: the use of the past tense. Thus, history is not just what "the triumph of the West" *was*: it is what it *still is* for the rest of the planet. There are examples too recent to need evoking: for some time there has been a tendency to say North instead of West, which makes the other countries, for binary reasons, the South. As in the previous case, there is no point in sticking to geographical origins. What we have are structural, not topographical, differences. I still think it preferable to remain with the "underdeveloping"/underdeveloped duality, which retains the colonizing traits of the difference. However, I have no wish either to adhere thus to a dichotomy of demons and angels. The colonizers have been wont to rely on intermediaries from the oppressed peoples for whom the division has been and still is very profitable. I use the words of someone who is indisputably a defender of our values, Darcy Ribeiro, from his book *La universidad nueva: un proyecto* [The New University: A Project], to which I shall be referring again:

> [T]he backwardness of Latin America is neither natural nor necessary, but it does exist and it persists because we have been complicit in its causal factors.... [I]t is not possible to dismiss the conclusion that the causes are within us, not in natural, innate or historical failings, but in forms of complicity of which we are guilty.... In fact, there is no doubt today that the project of colonial and neo-colonial exploitation of Latin America—a disaster for our peoples who paid the price in oppression, poverty and pain—was highly successful for those who dictated it and ruled as the dominant classes.... It was our class-based project of prosperity which led us, when colonial domination ended, to seek new forms of domination so that the old privileges could be maintained and extended.... All this sought to generate surpluses and sustain the royal prerogatives of a highly privileged social stratum in which academic intellectuals always found their place. (Ribeiro, 1973)

And further on,

> This is ... the project of the dominant-subordinate classes of Latin America who see in reflex modernization the only prospect of progress and prosperity compatible with perpetuation of their power and privileges. Faced with this threat, we must all choose between the role of modernizers and that of accelerators. That is to say, the role of reactionary joint actors of the innovative forces set on preserving the current social, economic and political system by means of

a merely modernizing transformation; or the opposing role, that of activists of the revolution required to remake the social order, and thus enable the potential of the new civilization to be realized for the benefit of the people as a whole.... The social revolution on the one hand and reflex modernization with all its regressive effects, on the other, are thus presented respectively as opposing options for the peoples and the ruling classes of Latin America. (Ribeiro, 1973)

I have no need to say which of these choices seems to me to be the only valid option to enable our Latin American civilization to achieve its consolidation. The allusion to Latin America is a reminder that such a civilization, forgotten or denied by so many authors who have looked so closely at the issue of civilizations, such as Spengler and Toynbee, is on the other hand recognized by a mediocre author who has plundered and watered down the former: Huntington (1996) for whom our civilization is "one of the nine civilizations existing in the world today." Commenting on the phenomenon, Walter Mignolo (2000) stated,

Leaving aside the fact that the categorizing logic of Huntington resembles that of the famous Chinese emperor mentioned by Jorge Luis Borges and adopted by Michel Foucault (1966) at the beginning of *Les mots et les choses* [The Order of Things], I would only like here to reflect upon the fact that Latin America is, for Huntington, a civilization in itself and not even part of the western hemisphere.... For Huntington, Latin America's identity distinguishes it from the West: "Although an offspring of European civilization, Latin America has evolved along a very different path from Europe and North America. It has a corporatist, authoritarian culture, which Europe has to a much lesser degree and North America not at all." (Mignolo, 2000, p. 46)

This led Mignolo (2000) to state,

It would seem that Huntington does not perceive Fascism and Nazism to be authoritarian. Nor does he see the fact that United States authoritarianism from 1945 onwards has been deployed in control of international relations in a new form of colonialism, colonialism without territoriality.

It was not necessary to await Huntington's confused theory (so well received by the conservatives in his country) for word of our civilization. Back in 1877, when he was barely 24 years old, Martí (1963a) wrote,

The natural and majestic work of American civilization having been interrupted by the Conquista, a strange people was formed with the coming of the Europeans: not Spanish, because new sap rejects the old body; not indigenous, because it has suffered the interference of a devastating civilization, two words which, being an antagonism, constitute a process; a people was created which was mestizo in form, which with the recovery of its freedom is developing and restoring its own soul. . . . All our work, of our robust America, will then inevitably bear the stamp of the conquistador civilization, but will improve and advance it, and will amaze with the energy and creative drive of an essentially different people, superior in noble ambitions and when wounded, undying. It is already living again! (Martí, 1963a).

Martí returned to the theme on many occasions and to other related ones, especially autochthony, and set them out in their final form in "Nuestra América" [Our America] (Martí, 1881), where, among other daring and enlightening ideas, he stated,

[I]n America the imported book has been conquered by the natural man. Natural men have conquered learned and artificial men. The native half-breed has conquered the exotic Creole. The struggle is not between civilization and barbarity, but between false erudition and Nature.[. . .] The European university must bow to the American university. The history of America, from the Incas to the present, must be taught in clear detail and to the letter, even if the archons of Greece are overlooked. Our Greece must take priority over the Greece which is not ours. We need it more. . . . Let the world be grafted onto our republics, but the trunk must be our own. And let the vanquished pedant hold his tongue, for there are no lands in which a man may take greater pride than in our long-suffering American republics. (Martí, 1891)

As will be noted in these texts, and in others of his, Martí habitually used the expressions "America" and "American" to refer to "our America" and the human creatures belonging to it; which by itself is a declaration of principles if one takes into account how common it is (and already was in the author's time) to give such terms other meanings. Martí could be said to have turned America into our America. In fact for him, the United States of America was "European America," which referred to their Western condition. Although I could devote the rest of my chapter to commenting on this inexhaustible text, which I consider to be the most significant contribution to our thinking (and I am far from being the only one who does), I will just exhort you to read an essay on him that I came across only recently: "Nuestra América," by a Portuguese Sociologist de Sousa Santos (2001, 18, pp. 185–217).

When I was a boy, and even afterward if someone had told me that sticking a needle in my ear would get rid of pain in another part of my body, I would have regarded it as a tasteless joke. Western medicine, at whose breast I was fed—that is how thoroughly Westernized I was—had not yet admitted acupuncture; it was majestically unaware that there is more than one map in the human body just as it was unaware and, to some degree, remains unaware of much of the knowledge from regions of the world that it regarded, or continues to regard, as more or less barbaric.

Self-referential general knowledge was presented by the West as its exclusive heritage. The rest was, if not silence, a picturesque jumble. Even as prestigious a region as the Arab world, one to which humanity owes so much, was treated with disdain. Edward Said (1978), as we all know, contested that attitude in his classic work *Orientalism*, which had already been in print for 25 years when Said died in 2003, at the very moment when the Arab world, which has already endured the horrors of the Crusades and considerable colonial aggression, is once again in the sight of the West. It is not surprising that, in the wake of that book, Said (1993) published another major work *Culture and imperialism* at a time when in academic circles it was regarded as bad form to use the word "imperialism" (open, for sure, to various interpretations) or to consider culture in the light of imperialistic depredations.

I should like to recall another singular fact: when the Europeans arrived for the second time on the continent later to be called America, carrying the seeds of capitalism (the first arrival, that of the Vikings, a thousand years ago, had virtually no impact), the Mayas had already discovered the zero, which the Europeans never discovered on their own (the Arabs brought it, like so much else, from India). One cannot possibly imagine the subsequent development of the quintessential "hard science" of mathematics without the mysterious cipher "signifying nothing," of non-European origin.

Faced with such examples, does it not make sense to imagine that other cultures, other populations (including those who used to be called primitive, to which the term "first peoples" is now rightly preferred) or the underprivileged may possess knowledge that might be of benefit to humanity as a whole? The great Mexican Alfonso Reyes liked to cite an expression he heard, I believe, from a Spanish peasant: "We know everything among all of us." Furthermore, in his "Notas sobre la inteligencia Americana" [Notes on the American Intelligence] (1936), Reyes recalls that in the meeting where his text was being read out, he affirmed, like the Argentine philosopher Francisco Romero, that ours was a culture of synthesis, and that

> neither he nor I was properly understood by our European colleagues, who believed that we were referring to a summary or basic compendium of the European conquests. According to this superficial interpretation, the synthesis is an endpoint. On the contrary: synthesis, in this case, is a new point of departure, a significant structure that contains within itself something new. H_2O is not simply a combination of hydrogen and oxygen, but is—in addition—water. The amount 3 is not simply the sum of 1 plus 2 but is—in addition— something that is neither 1 nor 2. This capacity to both look outside the incoherent panorama of the world, and to establish objective structures that represent a step forward, flourishes in the fertile and fertilized soil of the American mind. Compared with the average American, the average European appears to be perpetually enclosed by some Great Wall of China, and irrevocably, like a

provincial mentality. So long as they do not realize this and accept it modestly, the Europeans will have failed to understand the Americans. It is not a question of vulgar comparisons between what might be superior or inferior in itself but of differing views of reality." (Reyes, 1960, p. 88, note)

In "Posición de América" [Position of America] (1942) he clarified that on that occasion "we were referring not only to the European tradition but to all human heritage" (Reyes, 1942, p. 265). A short time earlier, in "Esta hora del mundo" [This hour of the world] (1939) he stated that Western civilization "set against the expanse of history, is a mere chapter and by no means a goal." (Reyes, 1939; 1960, p. 237) He expanded on that in "El hombre y su morada" [Man and his dwelling] [original edition, 1943]:

> We are not even certain that the Western mode will be the dominant one in the future. To believe otherwise is to accept as definitive a short-term egocentric error; it means the perpetuation of those absurd imperialistic ideas seeing anything outside the familiar context as not really part of humanity but rather as some kind of vegetation or as a fauna of "natives" destined to be sacrificed. (Reyes, 1943; 1960, p. 282)

I should like to draw attention to the fact that in the citations from Reyes, as in the case of Martí, "América" and "americanos" means "our America" and its inhabitants.

In his memorable book published in 1940 and entitled *Contrapunteo cubano del tabaco y el azúcar* [Cuban counterpoint: Tobacco and sugar], of which we Cubans are justly proud, Fernando Ortiz (1940) coined a term that would be widely accepted: *transculturation* whose primary aim, which was surpassed in reality, was to replace the English term *acculturation*. Ortiz devoted many pages of his book to explaining his neologism, as applied to Cuba. I shall confine myself here to a few lines:

> All cultural encounters are like the genetic coupling of individuals: the child always has something of both its parents, but is also always different from each of them. As a whole, this process is a *transculturation*, and this term embraces all the phases of its parabola. (Ortiz, 1940, pp. 136–143)

It is worthwhile asking whether the "culture of synthesis" used by Reyes to describe our America is not the result of what Ortiz describes as *transculturation* in regard to Cuba. I would go even further: at present, it is possible that all cultures are synthetic and, therefore, transcultural. In Cuba, the people's religion is the Santería, a blending of European and African heritages. It has always struck me that the Santería is defined as a syncretic religion (or even belief) to distinguish it from religions regarded as more important and more homogenous, such as Catholicism. But what religion is more syncretic than

Catholicism? Or, on another level, what culture is more syncretic than Western culture?

We should therefore proclaim the diversity of our America (the idea that it is unified by language, religion and so forth is no more than an illusion), with its plurality of origins, ethnic groups, peoples, languages, religions, arts, and knowledge. That does not contradict the fact that, beginning in 1492, we were thrown together into a common history, linked in its turn with that of humanity as a whole. In his controversial book *La pensée métisse* [Hybrid thought], Serge Gruzinski maintains that the mixing of cultures and the mixing of races that ensued worldwide began in America, in the chaos following the Spanish conquest, and that that first mixing of races, in different forms, was a forerunner of those we are witnessing at the start of the third millennium. Apropos of Mário de Andrade's poem "I am a Tupi who plays the lute," Gruzinski (1999, p. 21) comment,

> One can be a Tupi—an Indian from Brazil—and play a European instrument as old and refined as the lute. Nothing is irreconcilable, nothing is incompatible, even if the combination is sometimes painful. . . . Just because the lute and the Tupis have different histories, that does not mean that they cannot be brought together by the poet's pen or in an Indian village administered by the Jesuits.

Today we are witnessing an encounter that is even more spectacular and promising than that between the Tupi and his lute: the encounter between Subcomandante Marcos, on behalf of the indigenous peoples of the Chiapas region, and the Internet. That has brought the world a series of heartfelt messages, lyrical at times and filled with humor at others, from those who, suffering from five centuries of oppression and possessing knowledge going back much further, are setting about striving for a better world. On January 1, 1994, the North American Free Trade Agreement (NAFTA) between Canada, Mexico, and the United States of America, which according to its most enthusiastic supporters should turn Mexico into a first world country, entered into force. Significantly, on that very same day the world learned of the existence of the Zapatista Army of National Liberation Army (EZLN), whose most widely known spokesperson is Subcomandante Marcos. At the end of that same year the neoliberal program, contested by the EZLN, entered into a violent crisis in Mexico. Two years later, the EZLN held in La Realidad, on its own ground, a continental American meeting "For humanity and against neoliberalism." The many and necessary indigenous movements that are emerging in America, while aspiring to the just recognition of their autonomy, are not urging an impossible return to the pre-Columbian past, but are challenging the neoliberal globalization and the privatizations that have been creating havoc in America and, evidently, elsewhere. It was therefore very stimulating to find, at a recent conference sponsored by the Latin American Council of Social Sciences (CLACSO) and held in Havana on, October 27–31, 2003, not only Latin American social scientists of the stature of the Argentine Atilio A. Boron; the Mexicans Pablo González Casanova, Ana Esther Ceceña, and Víctor Flores Olea; the Brazilians Francisco de Oliveira and Emir Sader; and the Venezuelan Edgardo Lander, among many others, and major non-Latin Americans thinkers such as the North American Noam Chomsky, the Egyptian Samir Amin, the Briton Perry Anderson, and the Belgian François Houtart, among many others, but also indigenous leaders such as the Bolivian Evo Morales and the Ecuadorian Blanca Chancoso, who have been playing such an outstanding role in their respective countries emulating the work of Rigoberta Menchú of Guatemala, winner of the richly deserved Nobel Prize for Peace in 1992. Recently we were again witness to some absurd accusations against indigenous peoples made by an allegedly intelligent author who, unfortunately, has been getting broad press coverage, and who has been making the erroneous, if not monstrous, claim that the growing indigenous presence in America represents a regression. Quite the opposite. More than a century ago Martí declared that if things did not go well with the Indians, they could not go well with our America. In the 1920s, José Carlos Mariátegui observed that the Indian question was a social rather than an ethnic issue and had to be resolved by the Indians themselves. That is what they are doing now. They have more than sufficient knowledge and drive for that purpose.

Even though I have already used up most of my time, there are certain questions I should not like to overlook. One concerns our universities. While far from being an expert in that field (although I have passed through several universities either as student or teacher), I can hardly avoid mentioning our universities since they are the actual or supposed creators and transmitters of knowledge. The first university in America was founded in Santo Domingo in 1530 and universities continued during the colonial period under the auspices of religious orders and with a narrowly scholastic mission. Once the colonial period ended, the universities underwent formal but not really functional reforms. Darcy Ribeiro (1973) has summarized their history thus:

> The historical development of the Latin American university paralleled that of our societies. During the colonial period, the university provided instruction for the clergy and to the educated elites. After the colonial period, it continued to play the same role, training intellectuals with anti-clerical and anti-royalist attitudes, but still loyal to the interests of the ruling classes. Evidently, during that period of transition, the university had to be modernized . . . to ensure broader educational opportunities so that the children of the upper classes, now more numerous, would have the status of an academic degree and to

confer degrees on the jurists, doctors, engineers and so forth needed to keep the system running and ensure the well-being of those same classes. To achieve that goal, it was enough to abandon the Spanish models of higher education and adopt a subordinate replica of the Napoleonic model of the professionalizing university. . . . The university thus managed, through a renovation parallel to the modernization of the socio-economic system, not only to resume its role as an institution essentially in support of the existing regime but also to dignify, with the endorsement of a new set of values, that collusion. (Ribeiro, 1973, p. 21)

Nevertheless, the mimetic, colonial, or neocolonial nature of those societies prevented their universities from attaining the knowledge their circumstances required. Hence Martí, even before openly stating in his programmatic essay "Our America" that the European university had to yield to the American university, had spoken of "bogus mentality that the rudimentary and false culture of the universities and the accents of history impose on the Latin American peoples over their natural intellect" (Martí, 1963b, p. 201). Among the efforts made to dismantle the structures of that university, providing as it did a "rudimentary and false culture," none had a greater impact or was more fruitful than the university reform movement that emerged in mid-1918 at the University of Córdoba in Argentina and in which Deodoro Roca played the leading role. His manifesto "From the Argentine youth of Córdoba to the free men of South America," published on June 21, 1918, rejected the "senile immobility," "bureaucratic university," "mediocre teaching," "concept of authority," and "mentality of routine and submission" (Kohan, 2000, pp. 37–38). The movement's impact was widespread not only in Argentina but also in many other countries of "our America" and it drew attention not only to university issues but also to their relationship to the most pressing issues in those different countries. That was how many of the universities in America became centers of significant political awareness and concern.

I can but recall a conversation I had with a colleague at Yale University when, between 1957 and 1958, I taught there after University of Havana where I worked was closed down during the tyrannical regime of Fulgencio Batista. My colleague failed to understand why the students, above all, as well as quite a number of teachers in Latin American universities became involved in public matters instead of just getting on with their university work. Of course he had no knowledge of such events as the 1918 Cordoba Reform, with its consequences in many Latin American countries, like mine; and he could not even imagine that, half a century after Cordoba, the 1968 events of university origin were going to take place that gave rise to so much talk regarding this country and especially this city, but which also took place in many other countries and caused bloodshed for instance in Tlatelolco,

Mexico. They even stirred the previously peaceful universities of the United States, with their famous sit-ins protesting against the terrible aggression that the U.S. government was inflicting on the people of Vietnam, until memorably defeated.

But for all its importance (and it *was* important), the reform initiated in Cordoba fell far short of achieving the new universities intended. It could not do so oblivious to the conditions in the respective countries. And in recent decades those countries were to experience an all-out assault on the part of neoliberalism, with its exaltation of the market, its privatization cult and its disdain for activities that are nonproductive in material terms. It is no surprise that one of the most notable and recent books on the higher education centers, whose title is ironically reminiscent of the Cordoba Reform, is called *Las universidades en América Latina: ¿reformadas o alteradas? La cosmética del poder financiero* [Universities in Latin America: Reformed or altered? The cosmetics of financial power] (Marcela Mollis, compiler, Buenos Aires, CLACSO, 2003). The matter has been taken up by groups of authors such as those involved in that book; in number 17, 1996/1997 of the *Revista Chilena de Humanidades* [Chilean humanities review], or several of those taking part in symposia organized by UNESCO (1998) and UNESCO/ICSU (1999); and, of course, by individual authors, among whom I am pleased to highlight Pablo González Casanova, formerly rector of the National Autonomous University of Mexico (UNAM), in his work *La universidad necesaria en el siglo XXI* [The university necessary in the 21st Century]. The general views guiding such scholars are not solely negative. I mean they do not just raise objections to the havoc caused by neoliberalism, market theology, or privatization but come up with the counterweights of democratization and solidarity as the essential alternative. A singular and recent example of the need to possess a different and truly new university system was offered in the Final Declaration of the International Meeting "*In Defence of Humanity*," which on October 24–25, 2003 brought together in Mexico City a distinguished group of "intellectuals from the sectors of academia, the media, culture, and social movements in diverse regions of the world . . . to reflect on the extremely grave situation facing the world today." Among other burning issues, the Declaration includes this:

> To propose the creation of an international university whose goal will be to bring together humanists, scientists and artists from around the world, to dedicate their knowledge specifically to education, research and cultural dissemination, with the aim of achieving peace and a more free and just world. This university will bring together all intellectuals who pursue these objectives from democratic and socialist anti-imperialist perspectives. It will strive to establish communities of dialogue, with the participation of intellectuals of so-called high culture and intellectuals organically linked to the social movements of our

time. It will be organized in the form of a network with autonomous campuses, whose members will cooperate on both a personal and [a] long-distance basis in common projects. (Final Declaration, 2003)

An unavoidable question is the harmful effect of "the New Global Vulgate," as it was termed by Bourdieu and Wacquant (2000, pp. 149–150), which, originating above all in U.S. universities, dishes out its agendas to the rest of the world. Those authors state,

> The dissemination of this New Global Vulgate . . . is the product of a really symbolic imperialism. Its effects are all the more powerful and pernicious when this imperialism is assumed not only by the supporters of the neoliberal revolution (who, under the guise of modernization, set out to rebuild the world making a clean sweep of the social and economic gains of a century of social struggle, which are now brushed off as archaisms and obstacles to the new nascent order) but also by cultural producers (researchers, writers, artists) and left-wing activists who, in their great majority, continue to think of themselves as progressives . . . Nowadays many topics that directly issue from intellectual confrontations linked to particularities and forms of particularism of United States society and universities have imposed themselves, under apparently de-historicized guises, on the whole world. . . . It is . . . a United States discourse, although it is thought and given out as universal, and expresses the specific contradictions of the situation of academics who, cut off from any access to the public sphere and submitted to a great deal of differentiation in their professional environment, have no other terrain in which to invest their political libido than that of campus quarrels dressed up as conceptual epics. (Bourdieu and Wacquant, 2000, pp. 149–150)

If this was advanced by a French intellectual of the stature of Bourdieu, it will be understood that the situation of the universities, and of the general intellectual ambit, of our America is, where this question is concerned, far more serious. I obviously have no need to go into the case of the brain drain prompted, among other causes, by the material difficulties of many of the universities in Latin America (such as low pay and the poverty of their libraries and laboratories) and the wealth of U.S. universities. Let me focus on an assessment of their outputs. Thus, according to Abel Trigo (2003), "[W]e have tacitly accepted the never explicitly formulated and still less debated commonplace that Latin American cultural studies are an appendix, a substitute or a translation of supposedly universal cultural studies and in English." This is to forget, for example, "the work of critics and thinkers, cultural movements and trends of thought, institutions, publishing houses and publications that came about in Latin America contemporaneously . . . [and] provided continuity and renovated, in the passionate enthusiasm of the 1960s, a splendid cultural essay tradition." Trigo then mentions that in that period we produced dependence theory; the critique of internal colonialism; liberation theology and philosophy; the pedagogy of the oppressed; the theatrical practices of collective creation and street theater; the new cinema and new song experiments; the extremely widespread renewal (I hate the use of the most unliterary term *boom*) of the narrative and also, I add, of poetry (Trigo, 2003, pp. 381–382).

Allow me now briefly to refer to Cuba, my country. I know that its mere name prompts the most varied of reactions. When in 1955 I came for the first time, as an enamored student, to this city, it was hard to come across people with any knowledge of Cuba apart from the mambo, which was then starting to become international, the Havana cigar (from the name of the capital), and perhaps its rum. Nowadays, that is, from 1959 onward, to mention Cuba is to risk a dispute. I have no thoughts of taking part in it. It suffices me to know that I have enough years behind me to have known neocolonial Cuba, bloody and corrupt, as evoked in films such as the second part of *The Godfather* or *Havana*; and also to have witnessed the huge efforts that, working from that reality, have been made to build an independent country with solidarity, health, education, and social justice, which in many areas admits of no comparison with any other country of our America, or of the entire third world, and has even achieved in such matters as infant mortality and life expectancy figures performances quite in line with those of the "underdeveloping" (that is, causing underdevelopment) countries. As regards education and scientific research, it is an entirely literate country that has set out to make higher education universal (Ministry of Education, 2003), increasing the number of its university institutions as far as the municipalities, and availing itself of the most modern methods and media of dissemination such as television, in which there is an educational program and a program on several channels called *Universidad para todos* (University for all). With respect to scientific research, Cuba has set up institutions of international acclaim such as the Genetic Engineering and Biotechnology Centre, the Molecular Immunology Centre, the Finlay Institute—Centre for Research and the Production of Sera and Vaccines—, the National Scientific Research Centre, the Dr Pedro Kourí Institute of Tropical Medicine (IPK), the International Neurological Restoration Centre (CIREN), and dozens more. When Pablo González Casanova received the 2003 International José Martí Prize, awarded by UNESCO, after a generous speech by Mr Koïchiro Matsuura, director-general of the organization, the demanding Mexican intellectual said in a message of thanks:

> His [Martí's] conduct links up style, thought and politics with incomparable ethical values, today shared by several million Cubans. It is impossible to think of another world possible without the lessons of Martí, the universal

master....Anyone thinking that I exaggerate should go to Cuba and see what that people together with its Government has achieved for the dissemination of culture and for elementary, secondary and higher education; for scientific research and the humanities; for health, justice, democracy with people's power for and with the people, and for a universal collective will for peace and brotherhood with all the other peoples of the world, including that of the United States.... The world will find its path to peace, education and life, and that path will undoubtedly go via Cuba. It will take in among its classics the thought and conduct of José Martí. [. . .] Furthermore, I should be grateful if you would hand over the cheque for the Prize to the Permanent Representative of Cuba, Ambassador Rolando López del Amo, asking him to be good enough to send it to his Government as a modest contribution to the Country-University project in which Cuba is engaged. (P. G. Gonzalez, 2003, p. 13)

It is bad taste to end up talking about oneself, and I shall not do that. On the contrary, I shall finish with a word about all. The West and the rest, North and South, "underdeveloping" countries (others) and underdeveloped countries, whatever they may be termed, are the major challenges ahead, and these are political, economic, military, and environmental challenges. The knowledge that has been and may be achieved must preserve humanity in the face of such challenges. While ignorance is a calamity, mere knowledge can come up with horrors, and wars bear this out amply. Bertolt Brecht (1968) wrote these lines: "Clasping to them the children, /the mothers watch the sky in terror / lest in it should appear the discoveries of the learned." I was but a few years old when certain "discoveries of the learned" rained ruin on London with the V2s and on Hiroshima and Nagasaki with atomic bombs. Curiously enough, I was the surprised witness of how, decades later, it was sought to celebrate such discoveries as triumphs of science. As has been often said, science without conscience is extremely dangerous. I shall not dwell here on other well-known facts such as the situation where, while humanity is growing apace (in the final four decades of the past century the world's population doubled), the number of the poor and very poor is constantly rising, and places like sub-Saharan Africa, which was the cradle of humanity, risks being its initial tomb. I confess that after the fall of the undesired Berlin Wall and the dismantling of the Soviet Union (which had been on the wane for some time), when the right was exultant and much of the left was turning right, I read with surprise and gratitude, together with too many banalities, the book that Jacques Derrida published in 1993 entitled *Spectres de Marx* [Specters of Marx] and in it, among many felicitous pages, this paragraph:

A "new international" is being sought through these crises of international law; it already denounces the limits of a discourse on human rights that will remain inadequate, sometimes hypocritical, and in any case formalistic and inconsistent with itself as long as the law of the market, the "foreign debt," the inequality of techno-scientific, military, and economic development maintain an effective inequality as monstrous as that which prevails today, to a greater extent than ever in the history of humanity. For it must be cried out, at a time when some have the audacity to neo-evangelize in the name of the ideal of a liberal democracy that has finally realized itself as the ideal of human history: never have violence, inequality, exclusion, famine, and thus economic oppression affected as many human beings in the history of the earth and of humanity. Instead of singing the advent of the ideal of liberal democracy and of the capitalist market in the euphoria of the end of history, instead of celebrating the "end of ideologies" and the end of the great emancipatory discourses, let us never neglect this obvious macroscopic fact, made up of innumerable singular sites of suffering: no degree of progress allows one to ignore that never before, in absolute figures, never have so many men, women, and children been subjugated, starved, or exterminated on the earth. (Derrida, 1993/1994, p. 85)

In a subsequent text, "Culture is in danger," Bourdieu was to postulate "a new internationalism," called by Pascale Casanova, in The Republic of Letters, the denationalized international of creators and Bourdieu (2002) had this explanation:

> This tradition of specific, truly cultural internationalism is radically opposed, despite appearances, to what is termed "globalization." This word, which functions as both a password and a watchword, is the justificatory mask of a policy seeking to universalize the particular interests and the particular tradition of the economically and politically dominant powers, especially the United States, and to extend worldwide the economic and cultural model most favourable to these powers, presenting it both as a norm, an ideal, and as something inevitable, a universal destiny, so as to secure universal acceptance or, at least, resignation.... Those who remain attached to this tradition of cultural internationalism—artists, writers, researchers, but also publishers, directors of galleries, critics, of all countries—should today rally round at a time when the forces of the economy, which tend through their logic to submit cultural production and dissemination to the law of immediate profit, are finding considerable new strength in the so-called liberalization policies that the economically and culturally dominant powers are seeking to lay down universally under the cloak of *globalization*. (Bourdieu, 2002)

Bourdieu was to be even more explicit on the point in another text on "The International of Intellectuals," subtitled "Science as a profession, politics as a commitment: For a new division of political work," in which he speaks of

> a union of intellectuals thinking universally who may put themselves above the boundaries of States and in particular above the North-South divide, for a universal participation in universal goods.... What is required is to abandon the

academic microcosm and enter into contact with the outside world, above all with trade unions, citizens' associations and politically active groups. The idea is not to make do with the both intimate and compelling conflicts—always slightly unreal—of the scholastic world and to forge an uncommon mix of talents: competence and commitment. (Bourdieu, 2003, Criterios, p. 165)

If humanity, for all its present perilous situation, comes out on top, as we earnestly hope, it will only manage to do so thanks to that "new International" mentioned by Derrida (1993), which will not be just an "International" of intellectuals, unless the latter, as Bourdieu postulated, in addition to putting themselves above the North-South divide (the "universal" and the "local" according to the terminology with which I work), abandon the academic microcosm and "enter into contact with the outside world, above all with trade unions, citizens' associations and politically active groups."

But has this not been happening since Chiapas and Seattle up to Genoa, Prague, Québec, Porto Alegre, Cancún, and so many other places in the South and the North that have seen mass-scale protests against neoliberal globalization, the meetings of the owners of the world, and such neocolonial sophistry as the Free Trade Area of the Americas (FTAA)? Have not the huge demonstrations against the foreseen and savage aggression against Iraq been pointing in the direction of this "International"?

It as yet has no formal shape or definite face. But those actions afford us confidence that humanity (which, as Martí proclaimed, "is our country") will not stand idly by or waste itself in often artificial tensions when what is at stake is the survival of "The phenomenon of man" as Teilhard de Chardin (1955) called it, and as has been exalted by Ernesto Cardenal (1989) in his *Cántico cósmico* [Cosmic canticle] in verses that fuse poetry and science, religion and politics, and sorrow and hope.

References and Works Consulted

Bourdieu, P. (2002) La cultura está en peligro. In *Criterios. Revista Internacional de Teoría de la Literatura y las Artes, Estética y Culturología*, 33, Cuarta Época. Traducción del francés por D. Navarro, pp. 369–370.

—— (2003) La Internacional de los Intelectuales: La Ciencia como profesión, la política como compromiso, por una nueva división del trabajo político. In *Criterios. Revista Internacional de Teoría de la Literatura y las Artes, Estética y Culturología*, 34, Cuarta Época. Traducción del alemán por D. Navarro, p. 165.

—— and Wacquant, Loïc (2000) La nueva Vulgata planetaria. In *Casa de las Américas*. La Habana, 219, abril-junio, pp. 149–150.

Brecht, B. (1968) "De las bibliotecas," "Catón de guerra alemán." In *Poemas y canciones*, trans. J. López Pacheco and V. Romano. Madrid: Alianza Editorial, p. 111.

Cardenal, E. (1989) Cántico Cósmico [Cosmic Canticle]. In *Editorial Nueva Nicaragua Managua*. U.S. edition 1993, trans. J. Lyons. Willimantic: Curbstone Press.

Casanova, P. G. (2001) La universidad necesaria en el siglo XXI [The university necessary in the 21st Century]. Mexico City: Era.

Casanova, P. G. (2003) Mensaje de agradecimiento...[Message of thanks]. *La Jiribilla de Papel* (10), October, p. 13.
Chilean Humanities Review (1996/1997) Revista Chilena de Humanidades. No. 17, Chile.
CLACSO (2003) Conference sponsored by the Latin American Council of Social Sciences, Havana, October 27–31.
—— (2003) Las universidades en América Latina: ¿reformadas o alteradas? La cosmética del poder financiero [Universities in Latin America: reformed or altered? The cosmetics of financial power]. Compiled by Marcela Mollis. Buenos Aires.
Cordoba Reform (1918) History of Education: Selected Moments of the 20th Century. In D. Schugurensky (ed.). Department of Adult Education, Community Development and Counselling Psychology, OISE/UT.
Derrida, J. (1993) *Specters of Marx: The State of the Debt, the Work of Mourning, and the New International.* English translation by Peggy Kamuf and with an introduction by Bernd Magnus and Stephen Cullenberg. New York and London: Routledge, 1994.
De Sousa Santos, Boaventura (2001) Nuestra America: Reinventing a Subaltern Paradigm of Recognition and Redistribution. *Theory, Culture and Society* 18, pp. 185–217.
Elson, J. (1992) The Millennium of Discovery. Beyond the Year 2000. What to Expect in the New Millennium. *Time Magazine*, special issue, October 15, 92, pp. 16–18.
Final Declaration (2003) International Meeting "In Defence of Humanity." Speech delivered by Bolivian indigenous leader Evo Morales in Spanish at the Conference "En Defensa de la Humanidad" [In defense of humanity], Siqueiros Cultural Polyforum, Mexico City, October 24.
Franco, J. (1988) Beyond Ethnocentrism: Gender, Power and the Third-World Intelligentsia. In C. Nelson and L. Grossberg (eds.) *Marxism and the Interpretation of Culture*, introduction by C. Nelson and L. Grossberg. Illinois: University of Illinois Press, pp. 503–504.
Gruzinski, S. (1999) *La pensée métisse* [Mestizo Thought]. Paris: Librairie Arthème Fayard, p. 21.
Huntington, S.P. (1996) *The Clash of Civilizations and the Remaking of World Order.* New York: Simon and Schuster.
Kohan, N. (2000) *De Ingenieros al Che. Ensayos sobre el marxismo argentino y latinoamericano* [From Ingenieros to Che Guevara. Essays on Argentine and Latin American Marxism]. Preface by M. Löwy. Buenos Aires: Editorial Biblio, pp. 37–38.
Mariátegui, J.C. (1969) El problema del indio. In *Siete ensayos de interpretación de la realidad peruana*. Havana: Casa de las Américas, pp. 23–32.
Martí, J. (1891) Nuestra América [Our America]. Havana, Cuba: Centro de Estudios Martianos, Instituto Cubano Del Libro, p. 201. Published originally in 1891 in The New York Illustrated Review, New York, January 1.
—— (1963) Los códigos nuevos [The new codes] (1877). In *Obras completas* [Complete works]. Vol. 7. Havana: Editorial Nacional de Cuba, p. 98.
—— (1963) Eloy Escobar. In *Obras Completas* [Complete works]. Vol. 8. Havana: Editorial Nacional de Cuba, p. 201.
Márquez, G. G. (1982) *The Solitude of Latin America.* Nobel Prize Lecture, Literature 1981–1990, Singapore: World Scientific Publishing.
Mignolo, W. (2000) La colonialidad a lo largo y a lo ancho: el hemisferio occidental en el horizonte colonial de la modernidad [Coloniality far and wide: The Western hemisphere on the colonial horizon of modernity]. In *La colonialidad del saber: eurocentrismo y ciencias sociales. Perspectives latinomericanas* (The coloniality of knowledge: Eurocentrism and the social sciences. Latin American Perspectives). Compiled by E. Lander. Buenos Aires: CLACSO, Council of Social Sciences, Latin America, p. 78.

Ministry of Education (2003) Universalización de la educación superiorm [Universalization of higher education]. Report to the National Assembly of People's Power, Havana, Cuba.

NAFTA (1994) North America Free Trade Agreement (Canada, Mexico, USA). Creating the World's Largest Free Trade Area. NAFTA came into effect on January 1.

Ortiz, F. (1940) II. Del fenómeno social de la *transculturación* y de su importancia en Cuba [The social phenomenon of *transculturation* and its importance in Cuba]. In *Contrapunteo cubano del tabaco y el azúcar (Advertencia de sus contrastes agrarios, económicos, históricos y sociales, su etnografía y su transculturación)* [Notice concerning its agrarian, economic, historical and social contrasts, its ethnography and its transculturation]. Havana: Jesús Montero, pp. 136–143, last page citation.

Reyes, A. (1936) Notas sobre la inteligencia Americana [Notes on the American Intelligence]. Sur, Buenos Aires, September 1936. In *Obras completas*. Vol. XI. Mexico City: Fondo de Cultura Económica, 1960.

—— (1939) Esta hora del mundo [This hour of the world]. In *Obras completas*. Vol. XI. Mexico City: Fondo de Cultura Económica, 1960, p. 282.

—— (1942) Posición de América [Position of America]. In *Obras completas*. Vol. XI. Mexico City: Fondo de Cultura Económica, 1960.

—— (1943) El hombre y su morada [Man and his dwelling]. In *Obras completas*. Vol. XI. Mexico City: Fondo de Cultura Económica, 1960.

—— (1960) *Obras completas* [Complete works]. Vol. XI. Mexico City: Fondo de Cultura Económica, 1960, p. 88, note, p. 265, p. 237, and p. 282.

Ribeiro, D. (1973) *La universidad nueva: un proyecto* [The new university: A Project]. Buenos Aires: Editorial Ciencia Nueva, pp. 11–12, 16–17, and 21.

Roca, D. (1918) From the Argentine Youth of Córdoba to the Free Men of South America. Manifesto published on June 21.

Said, E. (1978) *Orientalism*. New York: Vintage Press.

—— (1993) *Culture and Imperialism*. New York: Knopf.

Teilhard de Chardin, P. (1955) *Le phénomène humain*. Paris: Éditions du Seuil.

Trigo, A. (2003) La larga *Marcha* hacia los estudios culturales latinoamericanos [The Long *March* towards Latin American cultural studies]. *Marcha* y América Latina [*Marcha* and Latin America]. Pittsburg: Horacio Machín and Mabel Moraña, Publishers, Instituto Internacional de Literatura Iberoamericana, pp. 381–382.

UNESCO (1998) World Declaration on Higher Education for the 21st Century. Adopted by the World Conference on Education, Paris, October 9.

UNESCO/ICSU (1999) Declaration on Science and the Use of Scientific Knowledge. Adopted by the World Conference on Science, July 1,1999. Budapest, Hungary, final text.

Zapatista Army of National Liberation Army (EZLN) (1996) Meeting for Humanity and against Neoliberalism. La Realidad [Ejército Zapatista de Liberación Nacional (EZLN)].

Chapter Seven
The Hybridization of Knowledge: Science and Local Knowledge in Support of Sustainable Development

Hebe Vessuri

The Power of the Idea of Science in Modern History

Many authors have repeatedly argued that in order to start solving the complicated problems of science and other forms of knowledge in the contemporary world, it is necessary to go beyond conventional claims to the universality of scientific knowledge and introduce into the analysis the dimensions of power and domination. The power to narrate, or to block the development or emergence of other narratives, has proved very important to culture and to political domination, and it is one of the main links between them (Said, 1994). Throughout its history, science has fashioned a very powerful narrative that has been instrumental in delegitimizing other descriptions of the world, while consolidating its own position as the preeminent form of knowledge.

In recent years, however, there has been a growing awareness in a number of fields, ranging from medicine to agriculture, that the modern world has paid a high price for its outright rejection of traditional and alternative practices (and the knowledge that underpins them), because of the "different" way in which they are expressed. Such knowledge usually comes into conflict with what is often referred to as the "modern scientific world-view," and has tended to be dismissed as little more than superstition. Its apparent lack of rational basis is seen as a reason to ignore it, without it being clearly realized that the criterion of rationality being applied is itself a cultural product of Western societies. In the current context where notions such as knowledge society and knowledge economy have become increasingly common, revising the idea of scientific knowledge in the light of different forms of knowledge may offer unexpected rewards.

The Great Narrative of Emancipation and Construction of Latin American Nations

An example drawn from Latin American history will serve to demonstrate the hold of some ideas and forms of discourse on the collective imagination, and their persistence over time (Vessuri, 2003). The great narrative of liberation and Enlightenment of the late eighteenth and early nineteenth century stirred into action not only Europeans fighting for new ideas of equality and human community but also those in the colonies who found inspiration in them and took up arms to cast off the imperial yoke. In Latin America, for instance, the aspiration to freedom and progress championed by a creative, enthusiastic group of politicians, intellectuals, naturalists, and artists fueled the fires of independence, giving to the proindependence movement a distinctly scientific tenor.

As a new historical reality, bound up with the promise of the New World, the presence of a large number of non-European races in the composition of the new Latin American nations conditioned social and cultural life from the early days of Latin American independence in the early nineteenth century. Political elites in the liberation movement attempted first to create integrated nations that would incorporate the indigenous populations socially, economically, and politically into the general body of civil society. This contrasted with the policy of separation and segregation practiced between the Amerindian population (the Indians' republic) and the dominant Spanish culture (the Spaniards' republic) that had been upheld (with only partial success) by the Spanish crown during the greater part of the colonial period. The process of integration of the indigenous population into the general body of civil society in the new republics took two main forms. Initially, under the clear inspiration of Enlightenment thinking, lip service was paid through legislative pronouncements to considerations regarding the different roots of the new nations, and the political inequality of the natives. With the passing of time, however, the rhetoric of symbolic integration was tempered, while economic integration was stepped up, as indigenous communal land was transformed into individual property, and these ethnic groups and their cultures started to be looked down upon and subjected to discrimination.

The new Creole nations of Spanish America came to regard themselves as being radically different not only from the local ethnics but also from the colonial peoples of Asia and Africa; heirs to the transplanted European settler communities, they often had difficulties to disengage from a sense of belonging to the culture of the "mother country." The same feeling was encountered in Brazil, where such expressions as "the civilized world," "civilized nations," "civilization," and "the Enlightenment" were frequently used, and reference to the European countries was considered legitimate both to elucidate national problems and to justify legislative proposals (Carvalho, 1988, p. 114). In the second half of the nineteenth century, Latin American elites readily saw themselves as the children of the Europe of the conquistadors, as part of the Christian world. As such, they felt no sympathy for the victims of the Conquest, even though this did not stop them from condemning colonialism—as the most progressive minds were doing at the time in the colonizing countries.

Halperin (1980, p. LXXXII) has suggested that because of the vulnerability of the new Latin American nations in the "civilized" international order of the second half of the nineteenth century, they felt that they had to make haste to narrow the gap between them and the more developed European countries. Political sovereignty would be defended by the Latin Americans with a zeal that reflected their conviction that international relations, particularly between the big powerful nations and the weak states emerging in outlying regions, contained a dangerous element of real or potential hostility, which was in any case inescapable. A need was felt to constantly take care, lest the international standing of the new states be diminished and they be treated as the "barbarian" kingdoms of other regions. So it was that, instead of nations built from a fruitful combination of native roots with European and other ethnic stock, the nations that emerged sought to identify with Europe, in many ways denying their hybrid mestizo character. Among other things, they denied and forcefully tried to blot out the knowledge traditions incorporated in ethnic cultures, whose fragmentary legacy has, in recent years, only just started to be recovered based on what remains of the archaeological, anthropological, and ethnological record.

Confrontation, Subsumption, and Obliteration of Knowledge Forms

As in the case of the rest of the regional historiography, the dominant narrative in the history of Latin American science has been that of the transplantation and adaptation of Europeans knowledge and technics by cultural activists and entrepreneurs. Both Europeans and Latin Americans were directly or indirectly motivated by economic demands for the harnessing of natural resources, political requirements for domination and colonial security, and, later in the nineteenth century, by the expansionist designs of European capitalism. Irrespective of their capacities, these scientifically minded intellectuals were "peripheral" in three ways: (a) by their marginal positions on the confines of European culture; (b) by their partial commitment to the scientific undertaking (driven by immediate imperatives of survival in often unstable environments, and the pressing economic and political needs of the new nations); and (c) by their roles as agents in tapping these natural resources of economic interest to the European centers of power, which were the source of sanction and reward. In several Latin American countries there were indigenous cultures that resisted the introduction of European ideas, artefacts, and technologies, such as in Mexico, Peru, Ecuador, and Bolivia. Sometimes the combination of cultural resistance and political persuasion led to open confrontation. In other cases, indigenous peoples were defeated by force and were pushed aside or decimated, as in Chile, Colombia, Venezuela, and Argentina, where perhaps the fiercest genocide occurred. Rather swiftly, then, wherever there existed some other "nonscientific" model of knowledge, it thus came to be divested of its legitimacy or power.

Peru offers one of the most promising avenues of research on issues of cultural hybridism in respect of the exchange of scientific and technological knowledge. In Peru, the indigenous majority again experienced growth during a good part of the nineteenth century—the only time in Andean history when cultural and demographic miscegenation ceased, if it did not go into reverse. In the early nineteenth century, the dominant white society weakened because of tensions linked to economic decline, political chaos, and institutional uncertainty in the process of colonial transition. There was a total lack of governance in rural Peru until at least the 1860s. Isolated and protected by the breakdown of national politics, the cities and markets of the indigenous communities were left to themselves. Cultivation of crops and other indigenous productive activities greatly increased through the development of local and regional circuits, in inverse proportion to the decline of Spanish markets for mining operations and for businesses situated in the urban centers. Throughout that century, the root of the Peruvian criollos' complaint about the "lack of workforce" for the coastal plantations or for guano mining was more social than quantitative, given that the rural populations were steadily growing.

The ability of the Peruvian ethnic groups to retain some linguistic autonomy suggests that they maintained strategic relations with the wider society and economy. They were surrounded without being overwhelmed by a hybrid Spanish-speaking society. At least three-fifths of the Peruvian population in the nineteenth century did not speak Castilian. The verbal contacts required for transactions were conducted principally in Quechua, although this language never became the lingua franca of the republic, unlike the indigenous languages in Bolivia, Ecuador, and Paraguay. The Peruvian peasants, while being partially integrated into the market, retained their language barriers, and revealed a deep social and cultural autonomy (Deustua, 1986, 1994). Nevertheless, a set of impersonal forces—market pressures, liberal ideology, and social classes—gradually began to influence indigenous attitudes, lifestyles, and social structures more and more widely, inevitably resulting in increasing integration. Thus it was the nineteenth century that was notable not only for the interruption of the centuries-old erosion of indigenous society, but also as an interregnum during which there was a shift toward new modes of ethnic assimilation that only recently has began to receive attention (Gootenberg, 1991).

There is much to be learned about the science and technology that speeded up the development of the export economy in the region from the end of the nineteenth century: the relation between agricultural productivity and changes in crop systems, the use of images of the natural world to create symbols of interethnic cohesion, the real impact of medicine and health on the demographic recovery of the indigenous population of Latin America, or the emergence of a scientific rhetoric in the nationalistic ideologies of the new countries (Cueto, 1995, p. 10). This view of science and technology was regularly put forward as being eminently "European" as far as it was "truly scientific." But more recently, the importance of ethnic factors in the social structure, a recognition of the experience and accumulated knowledge of indigenous groups, and strategies at the local and ethnic levels to counteract or compensate for the centralizing tendencies of the state from the precolonial period have begun to arouse the interest of researchers in studying the coexistence, tensions, complementarities, negotiations, and adjustments between popular and official knowledge.

What is clear from this example is that, contrary to conventional expectations, the bulk of the population of Peru seems to have recovered and prospered when the process of economic, political, and cultural hybridization was interrupted. We are thus led to think again about the simplistic view of cultural hybridization, usually seen as being positive in the sense of going hand in hand with an all-embracing and integrative one-way modernization in which modern science draws on other forms of knowledge as appropriate.

The world has changed since the nineteenth century, in ways that have surprised and often alarmed metropolitan Europeans who are now experiencing the presence of large nonwhite immigrant populations in their midst, and are faced with an impressive range of new voices with recently acquired power demanding that their own narratives be heard. These populations and voices have been around for some time, and they are a result of the process of globalization set in motion by modern capitalism. As Said reminded us, if we ignore or dismiss shared experience in the cultural field—where colonizers and colonized coexisted and struggled just as much through their representations as through competing geographies, narratives, and histories—we will miss out what was essential in the world in the past century (Said, 1994, pp. xvii–xviii). An important aspect was the mindset of the colonized. While an author like Fieldhouse refers to the British Empire, we endorse his argument as to a constantly renewed acceptance of subordination, extending it to the colonial cultural syndrome in general, whether through a positive feeling of shared interest with the mother country, or through an inability to imagine alternatives (cf. Fieldhouse, 1991, p. 103). This phenomenon was present in philanthropy, art, and science, notwithstanding the touches of disinterested collaboration that may have been apparent in particular circumstances. In considering the implications of this, we may go so far as to say that we cannot understand them if we leave out of the picture the dominant-dominated syndrome that persists to this very day, at the beginning of the twenty-first century.

In exchanges between Europeans and the "others" that began systematically 500 years ago in the capitalist era, the concept of identity at the heart of cultural thinking remained basically unchanged. The idea that scarcely varied was that there is a European "we" and "them" that is fairly stable, clear, and self-evident. From this standpoint, those regions of the world that were considered "remote" because of their distance from Europe did not have a life of their own, had no history or culture to talk about, and were without any independence or integrity worth considering without the presence of the West. Moreover, when there was something to describe, it was mystified either as a paradise on earth, or it was rejected as corrupt and irredentist. These images persist in the current process of globalization. The teleologies of modern times continue to point at a constantly receding target, leading many to wonder whether there is a goal ever to be reached at the end of the race. Is there indeed any end in sight? Meanwhile, some still seek to define independent goals in opposition to the hegemonic models of integration and imitation.

The Pairing of Science and Development

Since the Baconian manifesto in the seventeenth century, science has held out a promise of human well-being, although there have been increasing signs that the way in which science was advancing in modern times was creating as many if not more problems than those it was helping to solve. The boundaries between science and technology have become blurred in the past 200 years, despite the fact that once upon a time they could be narrated as separate histories. Even though today there prevails an image of science that renders it practically synonymous with technology, surprisingly the concept of pure science persists and there continues to be a conceptual difference between science and technology. For all that, the proportion of pure scientists in the world has dropped to less than 5 percent, and the numbers continue to decline. Why then is there still this concept of pure science when the pure scientist has now become a rarity? We believe that it is because not only is fundamental knowledge crucial to science but it also serves a useful social function, in that the social criticisms that are leveled against science can thus be deflected toward technology. For decades, scientific communities everywhere rejected pressure to change their rhetoric and define their research agendas in other terms than the exclusive quest for knowledge and truth. A fuzzy, ethereal concept of science having little in common with the real life undertakings of practicing scientists could thus be defended politically as the search for truth—uncontaminated by human greed, violence, and the quest for power.

In Latin America, as in the rest of the colonial and former colonial world, the political elites chose deliberately to see science as the state's responsibility while treating it as a sphere of knowledge that had to keep free of the restrictions of daily politics. In many Latin American countries, the state frequently decided to leave the practice of science outside the political arena, while seeing to it that the duly represented scientific community insured direct and privileged access to the state. In this "double vision," science became a justification for the state as well as a source of its domination. One of the distinctive paradoxes of the relationship between science and society in poor societies, which have little science and scant technological development of their own, is that the formula of keeping science outside the political arena served only to introduce another type of politics into science.

Accordingly, in countries like Argentina and Brazil there have been privileged sectors, such as nuclear power and/or electronics, and in others (e.g. Venezuela and Mexico) the same has been true of oil. In general, however, not only in sectors that were embedded in special institutions for reasons of state security, but also in the scientific and technological community as a whole, scientists were left relatively free of financial restrictions. Parliament approved their budgets as a matter of routine and without further scrutiny, and the spending of scientific laboratories and institutions was not subjected to public audit. Although universities (which in many countries had grown under the cover of modernization and development projects) have been declining in recent decades because of financial constraints, national political crises, and student population pressure, academic freedom has also meant, in the last analysis, the same lack of scrutiny and social accountability.

There is a feeling among scientists that they should enjoy a privileged position in society as experts and persons authorized to give opinions about culture, about what is right, or about the good life. Largely, the result is that what is recognized as right or as good is what coincides with the image of modern science, and what is bad or defective is what coincides with the things that obstruct standard modern science. The elites try to sell science as the cure for all the ills of Latin American society. The adaptation and copying of imported mature technologies are hailed as major advances in science. Of course, these assumptions about the perfect coincidence between the good life and science do not leave any room for other narratives or for the evaluation of the social contribution of the scientific establishment by ordinary citizens. They offer no possibility for the exercise of control over the scientific community, through a competitive political process and democratic participation.

There is a political asymmetry or inequality between scientists and the lay population, which is accentuated by the concept of "expert" that dominates the culture of modern science and allows lay people to criticize such science only in terms of its practical use, in other words its social and political utility—but not in terms of its ultimate value. In addition, through its instrumental-managerial orientation, modern science has established a strong link to development philosophy and practice. However, development specialists are coming up against the obvious fact that resources do not allow the consumer levels of the advanced countries of the North to be attained everywhere. In those circumstances, they may fall prey to a special interest, contributing to create an illusion of spectacular development consisting essentially in occasional demonstrations of technological capacity based on the standard model of technology transfer. With this model, highly visible small-scale technological achievements yield big political dividends in a particular context. Criticism is even allowed, in fragmentary fashion, although the very nature of technology as such is not called into question (Nandy, 1988).

That being said, analysis of attempts to address poverty alleviation reveals that development actions have often failed. Even worse is the fact that they may at times have had a harmful effect on peoples' ways of life. Solutions based on modern science have been used in situations where indigenous or local knowledge could have offered a better response, but for some reason or other, this was not even considered as an option (Sen, 1999). In the world where we live, the intellectual challenge is to offer a better understanding of how science and technology are gradually turning into a substitute for politics in many societies, and the need for a repoliticization of science based on a public audit, including persons outside the scientific community and its victims. This is not a popular idea among scientists, who form a rather privileged group, or in the urban middle class, hostile to the idea of politics and steeped in a culture whose hallmark is the acritical consumption of the products of science.

Science and Technology for Sustainable Development

Through the ages, the mode of knowledge production based on experimental control of research conditions that we now know as formal science, grew and became dominant at the expense of its relevance to concrete and complex situations. However, it is increasingly obvious that the continued functioning of science and technology, along the same lines as in the past, is not sustainable for the present and future well-being of the world and its inhabitants. Large swathes of contemporary human society are increasingly vulnerable to the pressures of technological change, much of which is of a global nature. Social, economic, and political inequality is reaching unprecedented heights, to a point where it threatens governance and the viability of even the prosperous portion of the world. The fact that members of the population have access to productive resources and keep them as their own preserve is of profound importance in determining patterns of economic growth and natural resource use. It is widely accepted that environmental destruction associated with the small production systems of small farmers is a consequence of their impoverishment, whether in absolute terms or in relation to other groups. However, much land has been degraded through the activities of individual and corporate interests (Bingswanger, 1991; Painter and Durham, 1995).

In such circumstances, the postulates of sustainable development are appealing as principles that can be reconciled with ethics and the conservation of nature. Its starting point is the reconciliation of society's development goals with the limits of the planet over the long term. This is a highly attractive policy even though it has proved difficult to define in precise terms (Parris and Kates, 2003, pp. 8068–8073). One of its basic postulates is that equal attention will have to be given to the ways in which social change shapes the environment through changes in demand, consumption, lifestyles, technological developments, political systems, and to the way in which environmental change shapes society, through changing ecosystem services for erosion, climate change, and so on. It means recognizing the fundamental complexity of interactions. This recognition of complexity also highlights the fact that an understanding of the individual components of nature-society systems is not sufficient to ensure an understanding of the performance of wide-ranging large-scale research programs, while at the same time it emphasizes regionalization and local specificity.

The current epistemological concern in science to address the challenges of sustainable development would appear to take up questions and redefine limits of inclusion and exclusion, interactions and openings that had been neglected since the nineteenth century in the boom years of experimental laboratory science. A number of increasingly interlinked movements springing from different contexts—including some scientific and technological circles—are seeking to redirect science and technology toward sustainable development. The prospect of the two endeavors eventually merging in a common design to produce human well-being seems at last to be in sight. Even the United States National Research Council, in 1999, gave it a name in the research and implementation program that has begun to grow out of these movements, referring to it as "sustainability science." No doubt, the strongest message to emerge from the Budapest World Conference on Science in 1999, and the Johannesburg Summit in 2002, was that the scientific research community needs to complement its historical role of identifying problems of sustainability by showing a greater determination to work on practical solutions to these problems (Clark and Dickson, 2003, pp. 8059–8061; UNESCO, 1999).

The transition to sustainable development is the latest reorientation of the development agenda. It is based on the idea of gearing science and technology to the need to stabilize the future world population, while alleviating hunger and poverty and maintaining the planet's life-support systems. This line of reasoning is clearly predicated on two main types of change. First, major global and regional changes to the climate and "health" of the biosphere have revealed as never before the need to secure the sustainability of ecosystems and ways of life. Second, the simultaneous processes set in motion by the globalized market in persons, ideas, and goods have opened up new challenges and new opportunities. The challenges are huge. In spite of the inherent difficulty in reaching a consensus as to what should be developed, what should be sustained, and for how long, it is also true that there exists a solidly documented and internationally negotiated consensus on issues relating to development and human well-being. There is far less consensus on paper around objectives and goals for maintaining life-support systems and living resources than for meeting human needs and reducing hunger and poverty (Parris and Kates, 2003). See figure 7.1.

An agenda directed toward problem solving does not mean reducing science to "applied" research. The search for practical solutions to the mounting challenges of sustainability makes it necessary to address a whole set of fundamental issues in particular fields of study, as is beginning to become clear in the context of emerging agendas in areas such as global environmental change, and around topics such as vulnerability and resilience, methodological approaches and monitoring technics, and so on. No doubt, notwithstanding statements to the contrary, there is always a risk that the most powerful stakeholders' agenda may come, this time as in the past, to dominate the setting of the science agenda. However, the need to reconcile the development goals of society with the environmental limits of the planet over the long term calls for skill in building a more human solidarity, founded on a stable society.

What likelihood is there that this new emerging program will refute the skeptical criticism of Nandy and other third world thinkers concerning the unending race toward moving, unreachable targets? Though the opposing forces are powerful, there is a small possibility that we can envisage the beginning of an answer to this question.

Science and Technology for Sustainable Development may be a strategic framework within which knowledge-related tensions become visible. In particular, this new approach entails an acceptance, and indeed a defence of

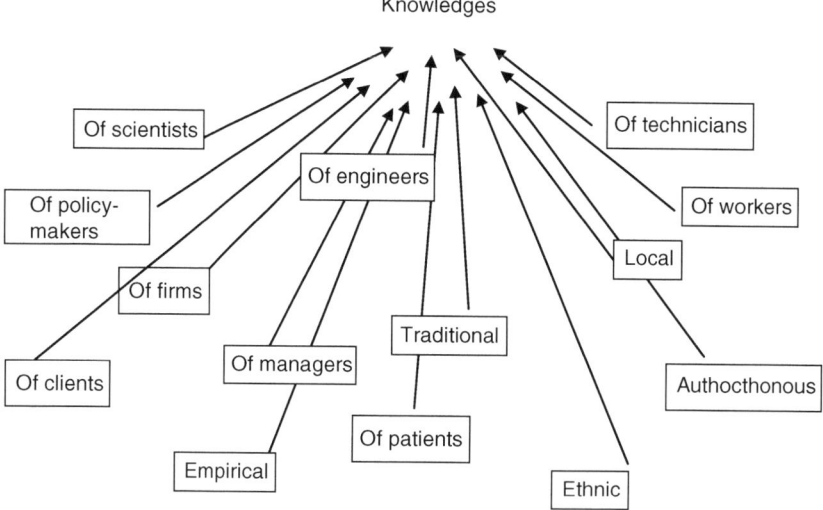

Figure 7.1 Forms of knowledge

heterogeneity. Science evolved historically by restricting what was heterogeneous in it and eliminating it, by way of social distancing and exclusion, denying its own rich heterogeneous sources. Such process of purification and separation between science and society conceals "hybrid monsters" (Latour, 1991) from our sight, which operate locally, mixing science and society. Instead of seeing them as aberrations, whose possible effectiveness has to be explained since they are supposed to occur only under special conditions, it is increasingly being admitted that they have always existed. Indeed, institutionalized and formal pure science may be considered an exception.

Distributed Knowledge

When Gibbons et al. (1994) wrote about "distributed knowledge," they were thinking of forms of knowledge and places for the production and reproduction of knowledge available in society outside academia, but they were still thinking basically in conventional societal terms when they referred to the knowledge possessed by managers, engineers, technicians, corporations, and so on. In fact, it may easily be argued that distributed knowledge also includes "other" forms of knowledge, local, traditional, empirical, ethnic, and so on. Nevertheless, the bulk of the population has remained beyond the effective reach of the concerns of science and does not share any part of the science agenda. One of the great challenges of S&T for sustainability is to open it up to include the poor and excluded, those who do not currently form part of the "public," those who do not have a voice. In the science and technology agenda for sustainable development, it is crucial to expand and/or redefine our view of the chain of knowledge production and consumption. What is at stake is the need to include other social actors, besides scientists and technologists in the strict sense at one end and mere consumers at the other.

There is a growing awareness that cultural diversity is a factor that can be effectively integrated into efforts to link up science and technology to sustainable development. Each culture faces challenges and requirements in accordance with its own values and customs. Scientists and technologists have the possibility of engaging in an open, constructive dialog with other groups in society that reflects the broad diversity of cultures and values in a world in search of new ways of understanding complex and interdependent aspects of sustainable development. In recent decades, lay citizens, particularly in connection with controversial topics of science and technology that have a clear public dimension, have steadily invaded the scientific and technological arena.

The examples of public health and environmental issues suggest that social movements may adopt distinctive forms of participation in scientific activities—even including "coproduction"—through close collaboration between practitioners and researchers. At the same time, their association with science and scientists significantly influences these movements. Given that different actors often expect different results from the application of S&T to problems of sustainability, systems of effective knowledge must offer scope for negotiation and mediation. Experience in the evaluation of scientific advice in general, and environmental and social evaluations in particular, suggest that scientific information can probably play an effective role in shaping social responses to public questions in so far as the information is perceived by the interested parties as being at once credible, important, and legitimate.

Particularly those who have suffered the consequences of scientific research need to be somehow included in the stakes of S&T research laboratories, in social research, in public health and legal studies, and so on. It is possible to discover real "experts" who are able not only to mediate between scientists and the lay public, but also to think critically about their experience and who may contribute usefully to the scientific process. Through learning the language of science, they may be able to translate their experience into a powerful criticism of the standard methodology of clinical proof, for example (Epstein, 1996). By framing their criticisms in such a way as to make them intelligible to scientists, they may urge them to respond (Collins and Pinch, 1998). The "lay expert" thereby contributes identifiable, partial, and useful knowledge, set in a critical context.

Hybrid Knowledge

The historical record of past centuries offers evidence of contacts, variations, transformations, exchanges, and instances of the transmission and learning of useful knowledge, making it difficult to maintain that there were no contacts between "indigenous," "local," "traditional" forms of knowledge and Western forms of knowledge (Agrawal, 1995, p. 422). Transfer and exchange have an integral role in the generation, growth, and dissemination of technology. There is thus a sense in which hybridity is a basic fact of knowledge. However, in striving to assert its purity and universality, science has used mechanisms of social distance and exclusion, denying its own rich, heterogeneous sources. A critical view of science in its current guise implies recognition and a reappraisal of several components, among them those actively directed toward the creation and strengthening of understanding based on new and old elements alike, which help to set in motion S&T work dynamics geared to ensuring sustainability.

The broad masses in Latin America and other parts of the poor portions of the world have been reduced to a negative example, a symbol of backwardness, and their creativity has been practically denied in most fields, whether technical or productive. Although looked down on by the Europeanized elites, the poor nevertheless continued to forge their fate throughout the nineteenth and twentieth centuries, with varying degrees of autonomy. The non-European roots of their cultures persisted more than expected, and occasionally their creativity blossomed within the confines of their officially unknown lives (Furtado, 1984). The impetus given to mass culture by the growth of the North American economy, supported by the new urban middle classes since the middle decades of the twentieth century, fueled consumerist values in many underdeveloped countries. Urbanization made the presence of the poor more visible; while it also became more difficult to deny their cultural creativity, with the result that the cultural frame of reference—based on the dichotomy between the elite and the poor—became shaky.

The emergence of an economically significant middle class introduced new elements into the cultural processes of developing countries in the context of modernization. The establishment of national R&D systems in the Latin American countries, linked to the growth of the middle classes, followed the principles and postulates of R&D in the developed countries and was closely associated with them, mainly through scientific research. For this reason, the needs of the popular majorities in developing countries did not turn into explicit demands to their R&D systems and were not generally regarded as topics of scientific research. This is at the origin of the incapacity of R&D systems in developing regions to solve the most pressing problems of their societies.

The combination of global, regional, and national crises in the past two decades has put an end to many of the illusions of the middle classes who, faced with unemployment and a sudden drop in income and standard of living, fear the possibility of a return to their social origins. The real way out of the impasse in which these societies find themselves does not lie in an attempt to reestablish privileges for a minority but in the adoption of economic, social, and cultural policies that tackle head-on the most pressing problems of the bulk of the population. This entails a recognition on the part of national R&D systems in developing countries of the existence of (a) technological bases that do not necessarily coincide with standard ideas; (b) a broad labor base that includes so-called "unskilled" (which does not necessarily mean "without knowledge") sectors of society; and (c) a revitalized interaction between technology and culture. In recent years there has been a notable change of attitude, so that where before the main concern was with the social actors serving as "gatekeepers," responsible for opening the way to modern technologies for development, now there is greater recognition of the importance of taking into account local technological know-how (Vessuri, 2001).

Sometimes these purveyors of local technology become actively involved in problem-solving and creative initiatives through learning-by-doing, and the loan, imitation, and transfer of formal science and technology; at other times such knowledge is simply removed from the social context in which it has a meaning, in order to be collected, analysed and classified. While in theory the purpose here is to safeguard it, this often has the effect of further weakening the position of the sources of local or indigenous knowledge. Those scientific disciplines that constitute stocks of knowledge developed by other cultures (as in the case of anthropology, geography, and ecology, for instance) are being remodeled in order to facilitate the recovery and better exploration of possible forms of development; local common sense is being reappraised and traditions (re)constructed. A complementarity may be envisaged between new kinds of technological "gatekeepers" and the purveyors of indigenous technology. In the new circumstances the former will act not as channels for an indiscriminate inflow of technology and culture from the developed world into their societies, but as agents who help to ensure the "discriminate and judicious" introduction of those technologies that are necessary (but not locally available), and equally the growth of indigenous technological capacity.

Hybridity in this context entails an attempt to find an answer within the limits of specific social circumstances to the many questions arising in the international context. The developing world's experience of "modernization" has been disheartening. The reproduction of modernized social structures through substitute industrialization has generally served to perpetuate technological dependency. Science, in turn, has played a part in reinforcing cultural subordination. People eventually became aware that the use of the surplus generated by international specialization to finance the consumption of a minority of the population made it possible to reduce the problem of resources but not to overcome the obstacle of the technological backwardness of the (poor) majority. Following repeated failed attempts at modernization, a poor country may end up with structural distortions that impede its subsequent process of development.

The Renewal of Science

In contrast with the reduced variability and heterogeneity that have characterized modern science, present-day environmental science, revitalized by modern ICTs, recognizes the value of accommodating the variation and heterogeneity of the sources of information and knowledge, giving renewed importance to questions of where and when. Hybridity does not only reflect the variety of interests involved but also recognition of the variety of expertise (already available both within official science and outside it in local communities of practitioners).

The Convention on Biological Diversity, for example, recognizes that there exists a significant body of knowledge on the history of ecosystem transformation and appropriate responses in local and traditional systems of knowledge. The Millennium Ecosystem Assessment (2003), for instance, considers that there is not much sense in excluding such knowledge simply because it is not "certified." Among the "practitioners" of environmental management, the private sector possesses substantial knowledge of ecosystems and policy measures, even though only a small fraction of this information is public. It is, therefore, important to effectively incorporate different types of knowledge. The challenge is to develop effective methods to judge the quality of the information.

During the cold war, modern science came to replace conventional politics when states began to compete through scientific projects redefined as spectacular technological breakthroughs. However, this was not new. The civilizing mission of colonialism in its encounter with the non-European world (often disqualified by Europeans, in the historical experience of contact with other peoples, as "savage") drew on a folklore composed of popular stories about colonial adventurers or explorers of a scientific bent who frightened or impressed the natives of Asia, Africa, or America with new forms of magic based on the discoveries of modern science. Recognition of the existence of other worlds and systems of knowledge offers the possibility of reversing the usual one-sided process by which modern science enriched itself through incorporating into it significant elements from other sciences—premodern, nonmodern, or postmodern—as further evidence of its universality and syncretism. Instead of using an edited version of modern science for Mexican, Peruvian, or Venezuelan ends, Latin American societies might come to use edited versions that would also encompass their traditional systems of knowledge for contemporary ends.

The call for a critique of the position of total dominance held by modern science is not aimed to revert to a pure and innocent, premodern world. It is, rather, a plea to include the point of view of marginalized peoples and cultures that have less and less of a say in the "expert" decisions that shape their lives. Not infrequently, they have to resort to the language of esoterism, mysticism, and life denial to withstand the brainwashing that would have them applaud each insensitive attack against their dignity, autonomy, and survival as a magnificent achievement of modern science. The challenge facing science and technology in the future is to contribute responsibly to sustainable development. For that, it needs to become truly universal generating knowledge that includes the contributions and concerns of the holders of traditional, local, alternative, indigenous, knowledge. This aim involves taking into account and building upon the ongoing struggles being pursued by the traditional knowledge systems that have survived. It means also valuing the systematic experience of those local people who have suffered the ill effects of modern technology in their immediate social context, and fighting against the exclusivist hegemony of modern science.

References and Works Consulted

Agrawal, A. (1995) Dismantling the Divide between Indigenous and Western Knowledge. *Development and Change* 26(3), pp. 413–39.

Bingswanger, H. P. (1991) Brazilian Policies that Discourage Deforestation in the Amazon. *World Development* 19(7), pp. 821–29.

Carvalho, J. M. de. (1988) Teatro de sombras: A política imperial. Rio de Janeiro, IUPERJ and Sâo Paulo, Vértice (Formaçâo do Brasil, 4), p. 196.

Clark, W., and N. M. Dickson (2003) Sustainability Science: The Emerging Research Program. In *Proceedings of the National Academy of Science* 100(14), July 8, pp. 8059–8061. Washington.

Collins, H., and T. Pinch (1998) *The Golem at Large. What You Should Know About Technology.* Cambridge: Cambridge University Press.

Cueto, M. (ed.) (1995) *Saberes andinos. Ciencia y tecnología en Bolivia, Ecuador y Perú.* Lima: Instituto de Estudios Peruanos.

Deustua, J. (1986) Producción minera y circulación monetaria en una economía andina: El Perú del siglo XIX. *Revista Andina* (8), pp. 319–378.

—— (1994) Mining Markets, Peasants, and Power in Nineteenth-Century Peru. *Latin American Research Review* 29(1), p. 29–54.

Epstein, S. (1996) *Impure Science: AIDS, Activism and the Politics of Knowledge.* Berkeley, LA and London: University of California Press.

Fieldhouse, D. K. (1991) *The Colonial Empires: A Comparative Survey from the Eighteenth Century.* Houndmills, 1965; reprint, London: Macmillan, 1982.

Furtado, C. (1984) *Cultura e desenvolvimento em época de crise.* Rio de Janeiro: Paz e Terra Economia.

Gibbons, M., C. Limoges, H. Nowotny, S. Schwartzman, P. Scott, and M.Trow (1994) *The New Production of Knowledge. The Dynamics of Science and Research in Contemporary Societies.* London: Thousand Oaks, and New Delhi: Sage Publications.

Gootenberg, P. (1991) Population and Ethnicity in Early Republican Peru. *Latin American Research Review* 26(3), pp. 109–57.

Halperín, T. (1980) Prólogo. Una nación para el desierto argentino. In *Proyecto y construcción de una nación (Argentina, 1846–1880).* Caracas: Biblioteca Ayacucho, pp. XI–CI.

Latour, B. (1991) *Nous n'avons jamais été modernes.* Paris: Éditions La Dévouverte.

Millennium Ecosystem Assessment (2003) *Ecosystems and Human Well-Being. A Framework for Assessment*. Washington, Covelo and London: Island Press.

Nandy, A. (1988) *Science, Hegemony and Violence: A Requiem for Modernity*. Delhi: Oxford University Press; new edition, Tokyo: United Nations University, 1998.

Painter, M., and W. E. Durham (1995) *The Social Causes of Environmental Destruction in Latin America*. Ann Arbor: University of Michigan Press.

Parris, T. M., and R. W. Kates (2003) Characterizing a Sustainability Transition: Goals, Targets, Trends, and Driving Forces. *PNAS* 100(14), Washington, July 8, pp. 8068–8073.

Said, E. (1994) *Culture and Imperialism*. London: Vintage.

Sen, A. (1999) *Development as Freedom*. Oxford: Oxford University Press.

UNESCO (1999) Declaration on Science and the Use of Scientific Knowledge. World Conference on Science, Budapest, UNESCO.

Vessuri, H. (2001) De la transferencia a la creatividad. Los papeles culturales de la ciencia en los países subdesarrollados. In A. Ibarra and J. A. López Cerezo (eds.) *Desafíos y tensiones actuales en ciencia, tecnología y sociedad*. Madrid: Biblioteca Nueva-OEI.

—— (2003) La ciencia en América Latina, 1820–1870. In G. Carrera Damas (ed.) *Historia General de América Latina*. Vol. VI. Madrid: UNESCO/Trotta.

Chapter Eight
Remarks on the Relationship between Knowledge Functions and the Role of the University

Akira Arimoto

Background

Before discussing the main topics of this paper, some considerations are needed to relate this work to the editors' overview.

First, there are reasons why the notions of "knowledge society" and "knowledge economy" need interrogation. Historically, rapid social change in connection with globalization and market mechanisms has brought about the strengthening of the relationship between knowledge and the economy in international perspective, under the pressure of an emerging knowledge society, and increasing international competition in the field of education—in addition to those of politics, economy, and culture.

Second, the case of Japan is interesting, as its response and take-off to join the rising knowledge society was delayed, mostly owing to the deep economic crisis caused by the breakdown of the bubble economy. Turning to a knowledge economy from a traditional knowledge base was, for many years, largely insufficient. Witness to this is the fact that the first formal usage of the term knowledge economy in the national government occurred as late as the end of the 1990s. Politically, especially in higher education policy, the Ministry of Education, Culture, Sports and Technology (MEXT), introduced, in 1991, a deregulation policy for the establishment of universities and colleges and, in addition, the University Council proposed, in 1998, the need for higher education reform toward the knowledge society and knowledge economy. In 2002, the national government proposed a series of higher education rationalization plans, including a university merger plan, top-down governance and management plan, corporatization of national university plan, and so on. In fact, the national government made the national university into a national university corporation in April 2004, and also started assessment of all universities in all sectors, including national, municipal, and private by introducing a nationwide new evaluation system.

Third, these facts reveal higher education reforms that inevitably have a great deal of impact on broad areas of academic work, consisting of teaching, research, service, and also governance, administration, and management. Since academia is expected to function as a key agency in the national science and research endeavor, it is natural that the so-called competition orientation principle has been introduced into academia, in order to obtain an internationally visible high competitiveness. To realize this purpose (in addition to the shift from a self-evaluation system to one of external evaluation), the previously used preevaluation system was drastically changed to a postevaluation system in the process of resource allocation among institutions.

Fourth, it is understandable that rapid reforms were taking place in Japan's economic, social, political, and higher education systems in accordance with worldwide trends of the knowledge society, knowledge economy, and globalization. It is especially noticeable that the connection between the knowledge society and the knowledge economy has brought about a strong expectation and effect on higher education, and, as a result, the importance of research and knowledge has been reconsidered to a considerable extent. In other words, its conditions, structure, and functions have been increasingly questioned, so as to promote high-quality academic productivity. It is not an oversimplification to say that both research productivity and teaching productivity, located at the core of academic productivity, have become key concepts in the nexus between knowledge society, knowledge economy, and higher education.

Fifth, when we place the present situation in a triangular relationship, between the nation-state, society (or market), and university, we see that the effect of the market upon the university has been much strengthened, and a demand-supply mechanism was also introduced in the process of encouraging competitiveness among institutions. On the other hand, it is also said that government control and expectation vis-à-vis the university has been strengthened, despite emerging market mechanisms in higher education and the government gesture to introduce deregulation orientation policies. The trend in this direction is observable in the example of control of the newly introduced national university corporation system as well as control of the evaluation process of all universities and colleges through the National Institute for Academic Degree (NIAD) and the Japan University Accreditation Association (JUAA). These kinds of policies, which are thought to be the first in 130 years in the history of higher education in Japan, are likely to strengthen the university's government control.

Sixth, these developments suggest that the national higher education policy today cannot ignore worldwide trends and phenomena in connection with the broadly emerging knowledge society, knowledge economy, and globalization. For example, the importance of quality assurance of higher education is necessarily and increasingly related to the functions of international agencies (such as UNESCO, OECD, WTO, World Bank, IMF, etc.), because these agencies have more or less strong relations in the process of elaborating a global standard of higher education.

Considering such background, the paper intends to make some arguments with regard to the theme "Knowledge Society Versus Knowledge Economy: Knowledge, Power and Politics," set up for the 2003 seminar of the UNESCO Forum on Higher Education, Research and Knowledge, focusing on the situation in Japan.

The main theme refers to the relationship between the function of knowledge and the role of the university as shown in the headline, paying attention to the background mentioned above, and the theme proposed at the UNESCO Forum seminar. Concretely, the paper is organized as follows: (a) knowledge functions; (b) the concept of academic productivity; (c) the relationship between knowledge and university; (d) the relationship between university and society; (e) research productivity; (f) teaching productivity; and (g) concluding remarks.

The connection of knowledge and other concepts is shown to be important in a series of topics: (a) knowledge and academic productivity, including research productivity and teaching productivity; (b) knowledge and university; and (c) knowledge and society. Discovery brings about knowledge differentiation to the extent that it makes general knowledge become advanced, specialized, and professional knowledge. Knowledge has several functions—such as understanding, discovery, dissemination, application, and control. In other words, these correspond to learning, research, teaching, service and administration, and management. Among these, research has an intimate relationship with knowledge's specialization and sophisticated differentiation, and this research function has been included within society as well as within the university.

Institutionalization of the research function into the university in modern society has brought about the concept as well as the function of academic productivity—especially research productivity, through which society has progressed considerably by way of academic and scientific development.

It is true to say that modern society, especially the emerging knowledge-based society, depends on academic productivity in order to realize its own development, and almost all nation-states opt to pay much more attention to such university activity to the extent that they attempt to build their own national and quasi-national universities, which exist between national and private universities as in state and public universities. Manifest and latent functions of knowledge are apparently affected by the process of social development: from agricultural, through industrial, to the growing knowledge-based society. In the modern university, a shift in priority of knowledge involvement—from a teaching orientation to a research orientation—has brought about conflicts between the research and teaching functions of knowledge, though the modern university is intended to integrate the two functions at least at a theoretical and ideological level.

Knowledge Functions

This qualitative approach, which stresses knowledge, is taken from the academic concern and tradition, mainly developed in the field of sociology of science, or sociology of knowledge. The approach focuses on knowledge, or academic work, assuming knowledge as the major determinant of university structure and operation. Such an approach emphasizes that academic work—such as learning, research, teaching, and service—is basically knowledge, or the application of knowledge both as stuff and means. In other words, we need to pay much attention to the nature of knowledge, of scientific knowledge and of academic disciplines.

The function of knowledge is mainly divided into understanding, discovery and invention, dissemination, application and control, as indicated in table 8.1. Accordingly, analysis of reforms occurring in the four phases—research, teaching, social service, and administration and management organization—is indispensable (Arimoto, 1996).

The Concept of Academic Productivity

Among the several knowledge functions, discovery is important since it brings about social progress by producing new knowledge through creativity and originality. In this context, the term scientific productivity—as well as academic productivity—used in the field of sociology of science, is still thought to be useful. Of course, the concept of productivity contains an economic tone, and its use suggests the possibility of some sort of invasion of the economic logic into academic world. Consideration of this kind of possibility may need a broad and generous understanding of the concept of academic

Table 8.1 Knowledge functions and their corresponding academic organization

Knowledge function	Equivalent	Academic organization
Understanding	Learning	Learning/study organization
Discovery	Research	Research organization
Dissemination	Teaching	Curriculum organization
Application	Social service	Social service organization
Control	Administration	Administration and management organization

productivity as a creative, original activity, academic vitality, and so on in higher education.

In the scientific community, the term scientific productivity was originally used by Merton (1938) in the sociology of science, focusing on the natural sciences as an indicator of the level of activity within the scientific community. Since the 1960s, some sociologists of higher education started to build a bridge between sociology of science and higher education research (cf. Becher, 1989; Ben-David, 1977; Clark, 1983). This trend was also observed in the cf. sociology of education in Japan: especially as one of the epoch-making incidents in this trend, the term academic productivity was introduced into the field of higher education research in Japan, in 1973, by Michiya Sinbori as a modified concept of scientific productivity—with a focus not only on the natural sciences but also on the humanities and social sciences (Arimoto, 1994; Shinbori, 1973). This concept was introduced into the sociological study of education in the author's original definition of this concept in the Shin-Kyoikushakaigaku Jiten (Japan Society of Educational Sociology, 1986, p. 5): "An indicator to know the creative activity outcome made by scientists involved in attempting to make new discoveries and inventions of social theory, law, concept, material, etc."

This new concept of academic productivity is still focused on research activity related to knowledge. In the present author's view, this concept is not only adaptable to research, but also to all functions of knowledge, and hence academic productivity is thought to apply to research, teaching, and service productivity. This concept is a total indicator of the level of activity of the academic community, while both scientific community and academic community share the concept of research productivity.

The stuff of such academic productivity is knowledge, including scientific knowledge, thought to be equivalent to academic disciplines.

The knowledge functions can be related to a productivity typology: discovery is related to research; dissemination to teaching; and application to service. In addition, control of knowledge can be related to governance, including administration and management. Accordingly, the main actors in academic productivity are defined by their knowledge functions: they are mainly faculty members, acting as researchers and scientists, teachers, and also consultants or administrators. Other incumbents of academia, such as nonacademic staff and students, also play certain roles, either manifestly or latently, in the processes of academic productivity.

The objects of productivity in research and teaching consist mainly of "eponymy" and "human resources." The former is perfectly adaptable to research productivity by the fact that remarkable discoveries are made in articles and papers published in academic journals. These discoveries are often evaluated and rewarded in the form of eponymy as in the Doppler Effect,

Boyle's law, and Newton's laws of motion. On the other hand, the latter is adaptable to teaching productivity, in which dissemination of knowledge is connected to production of human resources, which is valuable for social development. Teaching productivity means the products of teaching, which functions as one of the most important vehicles in academia. Teaching increasingly strengthens its importance in the emerging knowledge-based society, which in turn is substantially dependent on the dissemination of knowledge and demands higher-quality assurance of human resources.

The Relations Between Knowledge and the University: Transition from a Modern to a Twenty-First Century Higher Education System

As described above, the relationship between knowledge and the university is tight. This relationship, shown in figure 8.1, explains that today we are in the midst of a transition from the modern higher education system to the twenty-first century one. We can also see mutual relationships between society, government, and university in the modern higher education system.

As for the modern higher education (HE) system, including research, teaching, and service, it has among its knowledge functions the formation of the nation-state and society. These functions possess an almost closed structure within the university itself, and hence strong autonomy is given to the

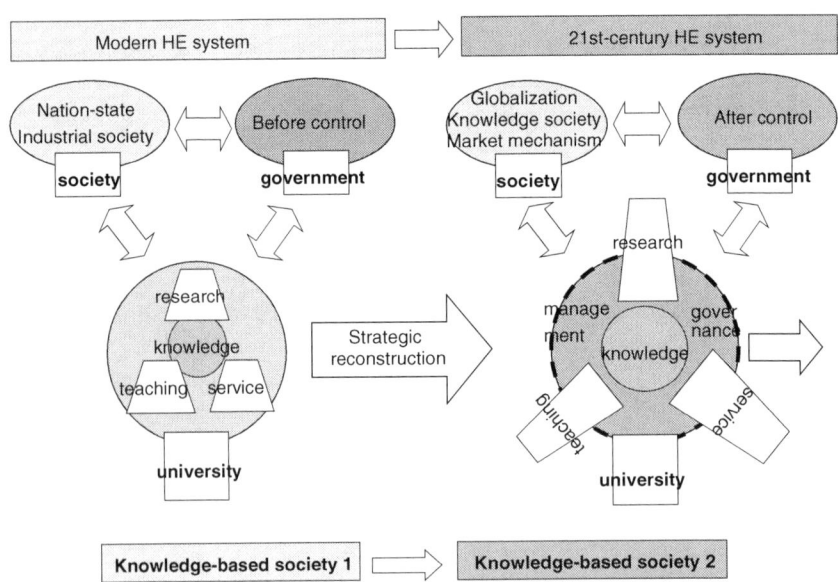

Figure 8.1 Knowledge, society, and university

university, which has a separate relationship with society. Accordingly, the university has the knowledge-based society integrated within it, although total society still remains separate from the academic world, being a less knowledge-based society. As we see later, we can call this stage a "knowledge-based society 1" (KBS1).

On the other hand, in the twenty-first higher education system, relationships between society, government, and university exist as well. It is understandable that society is inclined toward globalization, a knowledge-based society, and market orientation, while government is inclined to change from a before-control to after-assessment in the allocation of resources among universities and colleges. At this stage, university functions work not only in the university narrowly defined, but also in society at large. Hence, the knowledge-based society is not closed within the university, but it is opened to the total society. The borderline between university and society becomes borderless. This stage, when the total society also becomes a knowledge-based society, is called knowledge-based society 2 (KBS2), as we see later.

In the emerging KSB2, the present university is questioned both from inside and outside the university, especially from the outside, in pursuit of quality assurance and accountability of organization and system, with respect to academic work such as research, teaching, service, and administration, on the basis of knowledge functions. Many difficulties and conflicts are likely to occur in the shifting process from the modern university adaptable to KBS1, to the twenty-first century university adaptable to KBS2. Higher education reform is considered to be a positive action of pursuing a strategic reconstruction, so as to resolve such difficulties and conflicts with the aim of realizing the developmental shift of the modern higher education system to the twenty-first century higher education system.

As described above, academia's activity is based on knowledge. Expectation as to how the university uses knowledge seems to differ through the developmental stages, as a reflection of the effects of the society in which the university is located.

It is argued that in the medieval university, priority was put on knowledge dissemination, while in the modern university it is on discovery of knowledge. This was realized after the institutionalization of science in the university in the nineteenth century. Since then the major function of knowledge has been that of discovery. However, in the next stage, when the shifting from KBS1 to KBS2 takes place, dissemination is to be rethought, and understanding is to be stressed. This occurs because teaching and learning become more important due to massification and diversification of the student population, and also because society changes toward lifelong learning. Accordingly, it is observable that a core function of academia shifts from teaching in the medieval university, to research in modern university, and to learning in the future university.

Modern university, where research was institutionalized within its bosom, brought about some new characteristics: (a) differentiation of knowledge; (b) researcher's groups based on knowledge; and (c) organization of chair, department, faculty in which groups are inputs. As a result, the medieval university as a kind of academic community was transformed into the modern university as a kind of city and Ideopolis (Kerr, 1964, p. 93). It changes from "knowledge community" to "knowledge corporate." See figure 8.1.

As for the university organization, it shifted in modern society from "university" to "multi-versity" in Clark Kerr's conceptualization (Kerr, 1964) and in the future, it may shift to the "virtual university." In general, it is true to say that in the modern age, the university transformed itself from being an elite to a mass institution. The type of access to the university gradually shifted according to Martin Trow's prediction, from elite to mass and to universal stage (Trow, 1974). In accordance with it, major actors also change from teachers in the medieval university to researchers in the modern university, and to students in the future university. Respect and support of students in the teaching and learning process is becoming the focal point. The style of teaching, for example, is likely to move from lecture to seminar and personal instruction.

The knowledge function of governance is seen to be changing from control to deregulation, or privatization, throughout the world (cf. Amaral et al., 2002, 2003). This is also the case in Japan, though privatization here has already been realized to the extent that the private sector's portion of institutions and students is almost 75 percent. The result of this trend is shown in the shift of the national university to the national university corporation. The level of administration and management within the university as a knowledge function of control is also compelled to change as the reflection of social change. The method of management and administration shifts from a guild type of control to a rector type of control, as seen in European universities, and to a president type of control, as seen in American universities. In the Japanese university, it is changing from the former to the latter type by recommendation of the University Council proposal in 1998. As part of the same proposal, assessment as the outcome of the knowledge function, assessment methodology, and assessment indicators are gradually improved: it means development from chartering to accreditation, and to a mixture type. The dominant methodology of assessment and evaluation shifts from peer review to nonpeer review or, if we see it from a perspective of the method of academia, it shifts from "chartering" in traditional European universities to "accreditation" in modern American universities, and finally to a mixed type in the future.

If we observe the trend of relationship between social change and university development, there is no doubt that the university, which not only deals with knowledge as material but also has a close relationship to knowledge functions, has persisted for a long time since its foundation as the place of inquiry (Clark, 1995). At the same time, it is true that the university has changed in accordance with social change. The most common type of university shifted from the European continental type, involving collegiate universities developed in agricultural society, to the German university that has become a model for undergraduate programs as well as the American university, which has become a model for graduate study in industrial society, and finally to a new innovative university type including the virtual university in the knowledge society (Arimoto and Ehara, 1996). In the Japanese university, for example, mixtures of these models are competing with each other within both the level of the academic system and that of institutions: the German model, the American model, and the Virtual model.

The Relationship between University and Society: Conditions of University Reform

The university as a place of inquiry is keenly related to social development. Social development defines the university, and vice versa. The former is the social condition of the university, and the latter is the social function of the university. As pointed out, university's nature has gradually changed in accordance with social development for eight centuries. The effects of the university on society and age are not necessarily the same, although they have substantially many similarities in the sense that they have the basic functions of research, teaching, and service. The social function of the university is differentiated due to social development. For example, the function of research was not developed sufficiently before the appearance of the modern university, especially of the German university, where it has been institutionalized so well that it has become the prevailing paradigm among the functions of the university. These kinds of difference resulted from different social conditions in agricultural society and industrial society.

On the other hand, in the emerging knowledge-based society, the function of research is expected to become more important because discovery and invention of knowledge are thought to be indispensable factors related to social creativity and innovation. Teaching, which focused on recitation and memory in classroom learning so as to catch up with the given standard in the stage of agricultural and industrial society, will be changed in the knowledge-based society, where the trend toward creative education and problem-solving-oriented learning will increase in weight and where, at the same time, the nexus between research, teaching, and learning will increase in importance.

In addition to this kind of general trend, we can also point out that the present university is required to change its value, culture, and climate, so as to respond to the pressure from the past and future society: the social conditions for university reform are initiated from the social environmental changes in present society—partly due to the effects from both past and future society, and partly due to the effects from many nation-states worldwide. Social conditions for university reforms are also initiated from the logic intrinsic to the knowledge functions on which university activity is based. In this context, the following six parts are to be observed in details based on the author's previous study (Arimoto, 2003).

Effects from Past Society

Present society faces an age of huge structural change, which affects (among others) the relationship between society and university. Within this framework, society has experienced three developmental stages thus far: an agriculture-based society, an industry-based society, and a new knowledge-based society. In accordance with these three stages, the university has also changed, and is still changing—from a medieval university, to a modern university, and currently to a future university. The change of society from an industrial orientation to a knowledge orientation implies that such social conditions make university change its traits from those of the modern university to those adequate to the future university.

Today the university exists with an intermediate structure between the residual form of the medieval university, with its traits of guild control, community, universitas, uni-versity, collegiality, privilege, elite higher education—and the modern university, with its traits such as board of trustees, mass production, multi-versity, seminar, right of access, and mass higher education. The future university will reflect its own characteristics such as layman control, networking, virtual learning, obligation to society, and universal higher education. Currently, society reflects in its characteristics a hybrid of both industrial society, defining the modern university, and knowledge-based society, defining the future university. Pressure from the past reflects the logic of industrial society—including industrialism and science and technology. Factors such as industrialization, urbanization, massification, mass production, and mass consumption are deeply rooted in industrialism, which stresses quantity, scale, rationalization, and concentration. The modern university has developed the accompanying relevant traits such as quantity, scale, massification, and involvement as a product of society's values and codes.

At the theoretical level in higher education research, the Trow model is typically applicable to explaining the development of the modern university by means of the concept of progression through a sequence of elite, mass, and universal stages on the basis of student population: an elite stage with less than 15 percent of the 18-year-old cohort in university enrollment; a mass stage with more than 15 percent and less than 50 percent; and a universal stage with more than 50 percent (Trow, 1974). This model's logic is clear in the sense that it quantitatively explains the developmental stages on a basis of population: because of this characteristic, it is easily adaptable to the modern university appropriate to industrial society. However, it is likely to be difficult to adapt to a future university in a knowledge-based society in which qualitative criteria are necessary to understand the pathology related to the conflict between quantitative and qualitative factors, typically caused at the massification stage—and still more the postmassification stage—of higher education development.

The Trow model is readily comprehensible because of its explanation based on quantity. Certainly, the logic of its linear progression of higher education development—from elite to mass, and from mass to universal—is rational, but it is necessary to examine whether the model actually explains the real situation. To advance a conclusion here: real worldwide systems do not follow the American pattern on which it is solely based. Elsewhere, systems have not developed linearly according to the trend realized in the United States; rather the forms of their development reflect the traditions, cultures, and climates intrinsic to individual systems. This is already apparent in the change from elite to mass higher education. It is even more evident in the transitional stage from mass to universal higher education, where development is not merely oriented to a linear development but also brings conflicts between quantity and quality that are implicit corollaries to the quantitative development of massification. To explain this unexpected phenomenon, a concept of postmassification needs to be inserted between the two stages of massification and universal higher education (Arimoto, 1998).

Effects from Future Society

It seems obvious to say that the effects of future society will inevitably change the characteristics of present universities. The knowledge-based society, accompanied by a more profound orientation toward knowledge, lifelong learning, IT, and market mechanisms, is well established throughout the world. In this knowledge-based society, the university cannot monopolize knowledge as it did in industrial society; previously it was possible to do so because of a gap of knowledge development between society and university. The university is a knowledge society since it is theoretically committed to knowledge as the commodity used in conducting its academic work; we can call this KBS1. On the other hand, total society itself is now absorbing the university in an emerging knowledge-based society; this we can call KBS2. The type of society characterized on the basis of knowledge as KBS2 is analytically distinguishable from KBS1 appropriate to the university (Arimoto, 2002, p. 127).

The university is by nature a knowledge-based association, an organization whose foundation is knowledge. Now, however, society in general is becoming, to an increasing extent, a knowledge-based association. One of the key terms used is the concept of knowledge-based society. Analytically, a tentative distinction can be made between what I define as knowledge-based society 1 (KBS1) and knowledge-based society 2 (KBS2). The former refers to the academic enterprise, whose main aim is the development of knowledge; the latter refers to society at large, which is increasingly including academic activities within its own functions and roles. Society has lagged many years behind academia in carrying out the three functions of research, teaching, and learning—based on the advancement of knowledge, the resources of knowledge, and the academic discipline. However, the importance of knowledge is now part of the social fabric, and the distinction between society in general (KBS2) and the knowledge-based society of the university (KBS1) has begun to blur.

Competition between and integration of the two societies is thought to be accelerated by the process of shifting from industrial society to knowledge society. This may be especially so for the shift from KBS1 to KBS2. Similarly, within a knowledge society a shift from "Mode 1 to Mode 2" as identified by Gibbons and others will become clearer (Gibbons et al., 1994): that is, from analog to digital, from formal knowledge to tacit knowledge, from academic science to industrialized science, and so forth. Competition is necessarily observed between collegiate universities, incorporating the 800-year history of the traditional university, and the virtual universities emerging rapidly in KBS2. In relation to the knowledge economy, which accompanies a trend to strengthening the linkage between knowledge and the economy, this conflict will be encouraged to the extent that networking of virtual universities is increasingly promoted throughout the world. Accordingly, reconstruction of the whole university system, including the structures and functions of individual institutions and organizations, becomes inevitable in every country.

Effects from an International Perspective

Against this vertical perspective, there is a horizontal perspective focused on the influence that foreign universities have exerted on Japanese universities. Currently, universities are confronted with new and emerging situations in the field of education, as well as with the dynamics of social change evident in the dynamics connecting the knowledge-based society, globalization, and market mechanisms. It is clear that a lack of response to these dynamics of change immediately results in losing international competition. In its connection to globalization, development of market mechanisms brings about the situation that an economic logic of demand and supply prevails in the international perspective, and a consequent free competition is apt to invade the fields of research and education. In this regard, as Robert Merton pointed out (Merton, 1973), it is clear that an ethos proper to academic study has prevailed in KBS1, where a competition for priority is considered to be normal. Competition in promoting academic productivity, for example, is conducted in the academic community as an expected institutional activity closely connected to the functions and roles of the university. A series of researches in the field of sociology of science identified the fact that, in the international academic community, the center of learning and its periphery were clearly defined, so that a scrap-and-build mechanism was effectively operating around centers of excellence (COEs) (cf. Arimoto, 1994).

In KBS2, in the twenty-first century, where an orientation to information and knowledge will be increasingly promoted throughout the world, national economic productivity, must possess a closer relationship with academic productivity, and hence this kind of competition is necessarily encouraged and promoted.

As witnessed by empirical studies of the phenomena of formation and relocation of centers of learning, and centers of excellence, and of "brain gain" and "brain drain," international competition for obtaining and allocating knowledge, information and human resources among systems has become an objective of national power, prestige, and leadership by the accumulation of human powers, materials, and money in such centers.

Similarly, the construction of a higher education system possessing international competitiveness has become an important objective for all countries. As a result, a new trend has been recognized in rethinking evaluation systems, selection of key institutions, and concentration of the allocation of resources among institutions, by methods such as performance funding. In Japan, there is an expectation that universities will equip the nation for change through development of creative human resources, advanced science and technology, and the formation of COEs.

The Ministry of Education, Culture, Sports and Technology (MEXT) issued in 2001 a statement of policy for higher education, the "Outline of the Structural Reforms for National Universities," stressing the new role assigned to them. Briefly it proposed (a) to promote mergers and integration of institutions so as to reduce the present 100 national universities to a much smaller number; (b) to establish the national universities as independent institutions with a managerial structure; and (c) to build a group of 30 universities, able to attain the highest international levels by introducing the principle of competition and involving a third-party evaluation system. This is the so-called Toyama Plan, or more recently the "21st Century COE Programme." Based on this plan, a first selection of 113 programs in 5 fields was made in 2002; in 2003, a further 133 programs were selected in 5 other fields. Also in 2003, a variation of the COE program aimed at establishing centers of teaching excellence was established under the title of Centres of Learning (COL) and recently Good Practice (GP), and has been conducted in both four-year and two-year colleges; 664 programs were submitted and 133 programs were successful at the final stage of the selection process (JSPS, 2003).

It is clear that these new policies have been set up in an attempt to convert the previously existing procedure of protecting all institutions on an equal basis, to one of selective funding, to encourage competition among all institutions, to raise institutions' international competitiveness, and their ability to participate in the increasingly open, worldwide market.

Effects from the Nation-State

If we consider the effect of the "social realm" on the university, it is seen that it has changed over time. During the period of the extended medieval university, the main social realm lay within the city; in the modern university, it lay within the nation-state; and for the future university it will be located in global society. As pointed out by Clark Kerr, the modern university has acquired the character of a national university, such as Italian, French, German, English, Scottish, American, or Japanese (Kerr, 1994). Among them, the Japanese type incorporated components from leading university models from the advanced Western countries—especially the German and American models—and developed a structure that provided esteem to the central government-funded national universities. They were located in a system comprising national, public, and private sectors; massification of the system was effected substantially by the private sector. In Japan, the cluster of the former imperial universities (*teikoku daigaku*) was placed at the top of a doubly stratified system of universities, colleges, and professional schools, and of the national, public, and private sectors (cf. Amano, 1993). After World War II, although the former imperial universities were nominally classified as the one category of "daigaku" (university) together with other national and private universities, these artificially and intentionally stratified hierarchies have been persistently and manifestly maintained for many years, until today, by the national government.

Burton Clark made an international comparison of higher education hierarchies, pointing out that the Japanese type could be located in the same category as the English and French types, which have pinnacles in Oxford and Cambridge, and the Grand Ècoles; while the American and Canadian systems belong to an intermediate type showing a less steep stratification; and the German and Italian systems are almost flat with lack of hierarchy (Clark, 1983). Generally, developing countries have been inclined to control their national systems by providing a few national universities with protection and high prestige in order to strengthen the ability of their national higher education system in the process of catching up with those in the advanced Western countries.

Effects from Economic and Political Pressures

It is natural in an industrial society that investment for economic growth includes investment in universities, so as to promote productivity in both research and education, and also in social development. In the knowledge-based society, KBS2, where the knowledge economy expands through the interaction of knowledge and economics, knowledge itself is apt to be manipulated in the international marketplace. Growth of the knowledge-based society and a knowledge-based economy clearly strengthen the effects of the knowledge economy throughout the world (Gumport, 2002). Universities formed more or less on the basis of knowledge are necessarily affected by such economic trends, so that university systems are increasingly defined, not only by the national economy but also by the global economy. Accordingly, all countries are concerned about the fact that the ranking of universities is determined by an international marketplace, as well as a national marketplace. From a global perspective, an economic judgment is inclined to adopt the concept of academic productivity related to academic work including research, teaching, and service, the supply of human resources (workers), the salaries of workers, and the allocation of resources. As a result, it is evident that a general trend of demands for rationalization of universities from perspectives such as efficiency, effectiveness, and accountability has recently been accelerating.

In Japan, for example, the economic arguments in a report of a special committee appointed to advise the prime minister in relation to economic growth, stressed the necessity of introducing a policy of rationalization for universities and of application of market principles in order to allow competitive allocation of resources among institutions. The opening of the age of allocation, hitherto delayed in Japan, based on evaluation of institutional and organizational achievements, coincides with the arrival of the age of management and selectivity in the cause of greater efficiency and effectiveness. Moreover, institutional quality is assessed not according to the standards of the national system, but at an international level. Arrival of orientation to international markets translates into implementation of global standards and quality assurance for universities and higher education generally.

Logic Intrinsic to Academic Work in the University

Expectation for university reform from within the university is likely to be much more important than from outside, because it is derived from the academic logic intrinsic to the university. University reform is to be focused on responding adequately to knowledge functions. Among such functions, research and teaching are two indispensable functions. Within a university, emphasis is likely to be put more on research than on teaching, although both are expected to be integrated, because they are deemed to be the two wheels of the vehicle of academic work. Despite this, it is likely that an increasing separation of research and teaching provides a component in the breakdown of teaching. Rethinking the relationship between teaching and research has become a core problem in university reform.

The university enterprise is an organization properly dedicated to discovery and dissemination of knowledge. It is on the basis of knowledge that organizations and groups for research and teaching become differentiated vertically and horizontally in institutions of higher education. The fact that industrial society is a differentiated society is a reflection of professionalization based on the specialization of knowledge. Specialization of knowledge is typically developed in universities where separation of a series of activities generally consists of the substance of knowledge.

According to Clark's model, differentiation is developed on the basis of section, tier, sector, and hierarchy (Clark, 1983, 1995). The differentiation of faculty, department, and chair has developed corresponding to a section that has a horizontal differentiation of organizations and groups; the differentiation of undergraduate and graduate courses, and also that of associate, bachelor, master, doctor, and postdoctor has developed corresponding to the sector that has a vertical differentiation of organizations and groups. To the sector, understood as horizontal differentiation, belongs a social stratification of various kinds of institutions such as the research university, comprehensive university, professional university, liberal arts college, community college, and so on—while to the hierarchy, seen as vertical differentiation, belongs the social stratification of national, public, and private universities. These mechanisms, developed to maintain the university system, institutions, and organization, work mutually to produce conflicts between them. Any attempt to develop organizations and systems necessarily demands differentiation, so that the function of constantly coordinating conflicts arising from differentiation is an essential component.

Accordingly, rethinking the present function and role of the university both internally and externally is a requirement. As a result, there exist two sides to social condition and function.

The relationship between knowledge and university is tight to the extent that the university enterprise, which has been heavily concerned with material functions and the social contribution of knowledge, is an organization focusing (whether it realizes sufficiently or not) on the development of knowledge. It is because it gives adequate effect to learning, research, teaching, social service, and administration as knowledge functions that the university also brings about scientific and social development. Among the functions of knowledge, both research and teaching are especially important, and their organically integrated development is to be expected, though pressure toward their separation seems to have been working in the modern university as well as in society.

Academic Productivity in Research

Conditions of academic productivity have various facets according to the previous study (Arimoto, 1994). In the case of research, some processes are necessary to realize high research productivity: (a) a national policy for science and higher education development; (b) institutionalization of a scientific ethos into universities and colleges; (c) education and training of researchers in the process of scientific socialization; (d) evaluation and reward systems; and (e) the climate and atmosphere in departments and institutes.

Academic productivity might be effectively promoted if the scientific norm or ethos is internalized by academic staff. Robert Merton pointed out the existence of CUDOS (communality, universalism, disinterestedness, organized skepticism, competition, and originality) as the ethos of a scientific community (Merton, 1973). Such an ethos does not necessarily appear clearly among scientists, researchers, and scholars in contemporary universities and colleges that are becoming increasingly borderless with wider society— this is seen in the concept of KBS1 and KBS2 described above, and also in the emerging concept of Mode 1 and Mode 2 (Gibbons et al., 1994).

As far as research productivity is concerned, the relationship between the academic community within the university and human resources of researchers and students (as apprentice or embryo researchers) is sufficiently important to need intensive analysis. It is clear that, without such a relationship, neither development of human resources nor research productivity is likely to be available at all. As a result, it is true that so far, the advanced systems and institutions possessing centers of learning (COLs) or centers of excellence (COEs) have been successful enough in this direction; at the same time, peripheral systems and institutions need to be improved so that they can catch up with the COE groups in advanced systems. As described above, many bright students are in the process of brain drain from developing countries to developed countries.

In particular, the United States attracts students not only from developing but also from developed countries. For example, the proportions of foreign doctoral recipients (in all fields) from U.S. universities, who planned to stay in the United States in 1999, were large: from Europe, 40.8 percent; followed by

East/South Asia, 36.4 percent; North/South America, 30.2 percent; Pacific/Australasia, 28.6 percent; West Asia, 25.5 percent; Africa, 22.9 percent (National Science Foundation, 2002, pp. A2–50). It is correct to identify this as a symptom of globalization, and still more as an indicator of Americanization in the field of higher education and research.

If we consider the Japanese situation in relation to this, it is understandable that, as a case study of the developed countries, it is necessarily involved in dealing with many problems, in order to resolve the present situation. For example, the second proposal by the Human Resources Committee in the Council of Science and Technology is connected with a policy for training and preserving human resources for research to promote international competitiveness (Council of Science and Technology, 2003). A brief summary of their proposals is the following.

1. Arrangements for training human resources in the research function. Specifically this requires arrangements by those who are in charge of research human resources training—and those who are responsible for international competition—for introducing a sufficiently international research environment. For example, it means accepting top-level foreign researchers, collaboration with various overseas agencies in training human resources, constructing a research environment, including an environment for communication in English, with global standards. Training human resources by sending them to first-class organizations abroad is also necessary.

2. Realization of an environment in which a wide range of human resources are equipped to develop their abilities and commitment to research activity. Amongst others, the following practices are needed: construction of an open and fair personnel system in order to evaluate adequately the ability and achievement of researchers; and useful application of such results to the treatment of individual researchers.

3. Introduction of a supply mechanism able to cope with rapidly changing social demands for human resources for research. With this aim, it is necessary to construct a flexible training system that can accommodate social needs, and to establish partnerships between universities and industry. Promotion of recurrent education has to be established to support recruitment of human resources for research from other areas, both inside and outside the country, and for this purpose, the development of a model culture must be intensively encouraged.

Academic Productivity in Teaching

In the case of teaching, academic productivity means a commitment to teaching productivity; in turn, this leads to quality assurance of student ability and achievement as an outcome of teaching classes within the educational process. At the developmental stage of massification and postmassification in higher education, the average scores of student achievement are reported to have declined in some countries. Burton Clark, for example, identified this effect in the United States as the phenomenon of "schoolfication" of universities and colleges (Clark, 1997). Students have become diversified to the extent that they often lack the basic abilities and skills needed in reading, writing, calculating, and even thinking.

In an international survey on the academic profession in 14 countries, conducted by the Carnegie Foundation for the Advancement of Teaching, faculty members responded to a questionnaire related to student abilities. They identified declining achievement of students over a recent period of five years in almost all those countries where higher education had reached the massification stage (Altbach, 1996; Arimoto and Ehara, 1996). We should pay much attention to the implications of this with particular consideration of some of the specific problems: (a) the real situation of the process of student socialization in schools and colleges; (b) articulation of schools and colleges, and transition of students between the two educational segments, from a viewpoint of student development; and (c) especially, the teaching-learning process in colleges. In all aspects of input, throughput, and output of the teaching-learning process in universities and colleges, students as learners are considered to be central to the most important human resources in higher education, including presidents, vice-presidents, trustee committee members, faculty members, academic staff, nonacademic staff, and so on.

Improvement of teaching productivity in the stages of massification and postmassification depends on how the undesirable and insufficient state of student achievement can be raised to an international level by pursuing quality assurance of educational processes and outcomes. In order to realize this purpose and practice, a systematic improvement of input, throughput, and output—especially output—is necessary. Development of performance indicators (PIs) is inescapable, although with a few notable exceptions, such as Australia, it has not become established in the countries of the Asian and Pacific region.

As described above, there are many problems to be improved in the field of research productivity and many more in the field of teaching productivity. From an international comparison, the Japanese academic profession appeared to be committed to research more than to teaching, as shown in the result of Carnegie Foundation's survey previously cited. According to this survey, research orientation among faculty members in several countries is as follows: the Netherlands, 75.2 percent; Japan, 72.5 percent; Germany, 65.7 percent; Republic of Korea, 55.7 percent; United Kingdom, 55.6 percent; United States, 50.8 percent; Chile, 33.3 percent (Arimoto and Ehara, 1966). Contrary to this high proportion of faculty members with research orientation, the proportion of Japanese professors with a teaching orientation is only 27.5 percent, the second lowest following the Netherlands, with 24.8 percent. This type of high orientation to research and low orientation to teaching is recognized in Germany, Japan, the Netherlands, Republic of Korea, Sweden, and so on, while the type of high teaching orientation is recognized in South American countries (including Chile and Mexico), and also in Anglo-Saxon countries (including the United Kingdom, the United States, and Australia); Hong Kong and Russia also belong to this type.

If we see the relationship between research and teaching orientation productivity, there is an inclination toward the separation of the research orientation and the teaching orientation. How to resolve this situation is one of the important issues in modern universities and colleges throughout the world, and this is especially true in Japan where the weight of research orientation is very high. The ideal type is thought to be either teaching-centered integration or research-centered integration, as shown in table 8.2. According to Ernest Boyer's recommendation in his "Scholarship Reconsidered" (Boyer, 1990), it is clear that Japanese professors are categorized in type 2 (research orientation) rather than type 1 (teaching orientation) and, therefore, in the integration process of research and teaching under this kind of separation, type 4 (research-centered integration) are expected to shift to type 3 (teaching-centered integration).

In order to realize this kind of problem, higher education reforms are needed in terms of intentional integration of two separated orientations. Concretely, systematic improvement of output related to individual academic organizations, faculty, and students is necessary. Faculty development (FD) in particular is an important and identifiable activity for realizing this purpose and practice. The extent of institutionalization of FD into universities and colleges provides a kind of barometer to estimate realization of the purpose and practice of integration of two orientations (Arimoto, 2001). The United Kingdom and the United States are thought to be the most leading countries in the institutionalization of FD (at least from the point of view of certain Japanese researchers), since the early 1980s (Alstete, 2000; Arimoto, 2004).

In the case of Japan, the University Council (in 1998) proposed a necessity to accept at least a semiobligation of FD institutionally into university and colleges (University Council, 1998). Since then, faculty members' commitment to FD has been made to a considerable degree. However, quality assurance of FD activity is required at this initial stage of development, when its

Table 8.2 Integration of research and teaching orientations

Separation and integration	Teaching	Research
Separation	1. Teaching orientation	2. Research orientation
Integration	3. Teaching-centered integration	4. Research-centered integration

institutionalization has been improved to the extent that more than 60 percent of all institutions—and more than 90 percent of national institutions—have undertaken FD activity thus far.

In addition to this kind of systematic improvement of academic organizations, the culture and climate, more involved in research than teaching, should be intensively changed—so as to realize the integration and nexus of research and teaching, in accordance with Boyer's concept.

According to the recent nationwide survey of university presidents, the institutionalization of FD has improved much between 2003 and 1989. In particular, they agreed with the fact that faculty members were increasingly and earnestly committed in teaching improvement practices (Arimoto, 2004), with 81 percent agreement in 2003 compared to 55 percent in 1989.

Concluding Remarks

1. The main activity of the university is based on knowledge function has, since the origin of university, lasted until today, and is expected to remain so in the future, with a changing trend of society from KBS1 to KBS2. Roles of university, consisting of research, teaching, service, administration, and management, have a deep relationship with the full promotion of the function of knowledge.

2. The promotion of functions of knowledge is translated into the concept of "academic productivity," which has been developed from the field of the sociology of science. It was modified in the sociology of education to the concept of "scientific productivity," and dealt with in the humanities and social sciences, in addition to natural sciences. Moreover, the author of this paper thought of the theoretical possibility of academic productivity—including teaching, service and administration, and management productivity. Among these, he also pointed out the importance of both research productivity and teaching productivity paying much attention and consideration to detail.

3. Total society was faced with the advancement of the information orientation society in the 1960s and knowledge-based society in the 1990s, while university including functions of knowledge is thought to be knowledge-based society. The author called the latter KBS1, and the former, KBS2. The borderline between two societies, university and total society, is becoming ambiguous and borderless, since those societies are merging and, to a considerable degree, integrated. This trend has partially decreased and partially increased the value of the university's functions. At least high qualitative productivity in research, as well as teaching and education, will still contribute greatly to social development. Research productivity with high quality can contribute to the development of science and technology, and, through this process, it can contribute to social development. At the same time, teaching productivity with high quality can also contribute to the development of human resources, and through this process, it can contribute to the social development. In this context, every higher education system is to be necessarily involved in high academic productivity with focus on teaching as well as research productivity in emerging KBS2.

4. University reform substantially involved in such a kind of academic productivity is caused by various pressures from both inside and outside social environment to university, including external pressures from the past, present, future society, and the internal pressures from logic intrinsic to university.

5. It is necessary for us to make a study of the conditions of research productivity in order to enhance its quality and standard. It is also necessary for us to make study of conditions of teaching productivity, paying close attention to the condition that teaching has less priority than research in modern university. In addition to this intention, the nexus of knowledge functions, especially that of research, teaching, and learning, is to be stressed more in the emerging KBS2.

References and Works Consulted

Alstete, J. W. (2000) *Posttenure Faculty Development: Building a System of Faculty Improvement and Appreciation*. San Francisco: Jossey-Bass.

Altbach, P. G. (1996) *The International Academic Profession: Portraits of Fourteen Countries*. Princeton: Carnegie Foundation for the Advancement of Teaching.

Amano, I. (1993) *Kyusei Semon Gakko* [Old professional schools]. Tokyo: Tmagawa University Press.

Amaral, A., G. A. Jones, and B. Karseth (eds.) (2002) *Governing Higher Education: National Perspectives on Institutional Governance*. Dordrecht, Kluwer: Academic Publishers.

Amaral, A., V. L. Meek, and I. M. Larsen (eds.) (2003) *The Higher Education Management Revolution?* Dordrecht: Kluwer Academic Publishers.

Arimoto, A. (ed.) (1994) *Gakumon Chushinchi no Kenkyu: Sekai to Nihon nimiru Gakumontekiseisansei to sono Joken* [Study on the centres of learning: Academic productivity and its conditions in the world and Japan]. Tokyo: Toshindo Publishing Company.

—— (1996) Cross-National Comparative Study on the Post-Massification Stage of Higher Education. *Research in Higher Education* (25), pp. 1–22. Hiroshima: Research Institute for Higher Education, Hiroshima University.

—— (ed.) (1998) *Post Taishuka dankai no Daigaku Sosiki Henyo nikansuru Hikakukenkyu* [Comparative study of shift of academic organizations in post-massification stage]. *Higher Education Review* (46). Hiroshima: Research Institute for Higher Education, Hiroshima University.

—— (2001a) FD no Seidoka niokeru Shakaitekijoken no Yakuwari [Role of social condition in institutionalization of FD]. *Daigaku Ronshu* (31), pp. 1–16. Hiroshima: Research Institute for Higher Education, Hiroshima University.

—— (ed.) (2001b) *University Reforms and Academic Governance: Reports of the 2000 Three-Nation Workshop on Academic Governance*. RIHE International Publication Series, no. 7, March. Hiroshima: Research Institute for Higher Education, Hiroshima University.

—— (2002) Globalization and Higher Education Reforms: The Japanese Case. In U. Enders and O. Fulton (eds.) *Higher Education in a Globalising World*. Dordrecht: Kluwer Academic Publishers.

—— (2003) Recent Higher Education Reforms in Japan: Consideration of Social Conditions, Functions, and Structure. *Higher Education Forum*, 1, December, pp. 71–87. Hiroshima: Research Institute for Higher Education, Hiroshima University.

—— (ed.) (2004) *FD no Seidoka nikannsuru Kenkyu (1) 2003 gakucho Chousa Hokoku* [Study on institutionalization of FD (1): Report of 2003 Survey on University Presidents]. Hiroshima: Research Institute for Higher Education, Hiroshima University.

Arimoto, A., and T. Ehara (eds.) (1996) *Daigaku Kyojushoku no Kokusaihikaku* [International comparison of academic profession]. Tokyo: Tamagawa University Press.
Becher, T. (1989) *Academic Tribes and Territories*. London: Open University Press.
Ben-David, J. (1977) *Centres of Learning*. New York: McGraw-Hill.
Boyer, E. L. (1990) *Scholarship Reconsidered*. Princeton: Carnegie Foundation for the Advancement of Teaching.
Clark, B. R. (1983) *Higher Education System in Cross National Perspective*. Berkeley: University of California Press.
—— (1995) *Place of Inquiry*. Berkeley: University of California Press.
—— (1997) Small Worlds, Different Worlds: The Uniqueness and Troubles of American Academic Profession. DAEDARAS. *Journal of the American Academy of Arts and Sciences* 126(4), Fall.
Council of Science and Technology (2003) *Policy for Training and Reserve of Research Human Resources So As to Promote International Competitiveness* (in Japanese). Tokyo: Human Resources Committee, Council of Science and Technology.
Gibbons, M., H. Nowotny, C. Limoges, S. Schwartzman, P. Scott, and M. Trow (1994) *The New Production of Knowledge: The Dynamics of Science and Research in Contemporary Societies*. London: Sage Publications.
Gumport, P. (2002) Universities and Knowledge: Restructuring the City of Intellect. In S. Brint (ed.) *The Future of the City of Intellect: The Changing American University*. Stanford: Stanford University Press, pp. 47–81.
JSPS (Japan Society for the Promotion of Science) (2003) *Overview of the 21st Century COE Program*.
Japan Society of Educational Sociology (1986) *New Dictionary of Educational Sociology* (in Japanese). Tokyo: Toyokan Publishing Company.
Kerr, C. (1964) *The Uses of the University: The Godkin Lectures at Harvard University*. Cambridge: Harvard University Press.
—— (1994) *Higher Education Cannot Escape History: Issues for the Twenty-First Century*. New York: State University of New York.
Merton, R. K. (1938) *Science, Technology and Society in Seventeenth Century England*. New York: Howard Fertig Inc., 1970.
—— (1973) Falta Referencia. In N. Storer (ed.) *The Sociology of Science: Theoretical and Empirical Investigations*. Chicago: University of Chicago Press.
National Science Foundation (2002) *Science and Engineering Indicators 2002*. Vol. 2, appendix, tables. Arlington: National Science Board.
Shinbori, M. (1973) Study of Academic Productivity (in Japanese). In *Daigaku Ronshu*, no. 1. Hiroshima: Research Institute for Higher Education, Hiroshima University.
Trow, M. (1974) Problems in the Transition from Elite to Mass Higher Education. *Policy for Higher Education*. Paris: OECD.
University Council (1998) *21Seiki no Daigaku-zo to Kongo no Kaikakuhosaku nitsuite—Toshin* [University image of 21st century and reform planning ahead—proposal]. Tokyo: University Council.

Index

academic freedom, 45, 80, 84, 86–87, 89–95, 97, 163
academic productivity, 29, 176–79, 187, 189, 191–96
academic profession, 86, 93, 192–94
accelerators, 140
access to education, 9, 62, 72, 74, 86–90, 94, 97, 98, 99, 108–11, 114–17, 133, 182, 184
access to work, 117–22
accountability, 7–10, 12–13, 16, 53, 56, 71, 70–71, 97, 163, 181, 189
acupuncture, 142
Africa, 19–22, 80–82, 92–101, 102, 103, 127, 128, 133, 144, 151, 158, 171, 192
African higher education research, 73, 86, 92, 101, 103
African intellectual diaspora, 98, 99
African universities, 20, 69–74, 86, 90, 94–103
Agrawal, A., 169, 172
Aina, T. A., 82, 101, 103
Ake, C., 82, 103
Algeria, 26, 102, 108–23
Alstete, J. W., 194, 196
Altbach, P. G., 67, 77, 102, 105, 106, 193, 196
Amano, I., 188, 196
Amara, M., 136
Amaral, A., 31, 32, 182, 196
Americanization, 192
Americans, 143–45
Amin, K., 125–26, 135
Amin, S., 81, 103
Anderson, B. P., 146
Ang, I., 90, 103

Annan-Yao, E., 103
anti-imperialism, 143–46, 148–49
Arab Human Development Report, 27, 125, 127, 130
Arab women, status of, 27, 110, 120–23, 126–35
Arab world, 26, 27, 111, 122, 125, 126–28, 131–35, 143
Argentina, 147, 160, 163
Ariel's Alternatives, 139
Arimoto, A., 29, 178, 179, 183, 184, 185, 186, 187, 191, 193, 194, 195, 196, 197
Arnove, R. F., 78
Association of African Universities (AAU), 23, 100, 102
Association of Commonwealth Universities (ACU), 56–58, 60, 63–64, 69–70, 76, 77
Asturias, M. A., 138
Australia, 11, 90, 139, 193
autonomy, 13, 14–16, 18, 41, 45, 55, 87, 89, 97, 98, 145, 160, 169, 172, 180
autonomy-accountability tension, 16

Bachrach, P., 37, 49
Balintulo, M., 95, 103
Banerjee, H., 101, 103
Baratz, M., 37, 49
Bardach, E., 49
Barnett, R., 61, 75, 77, 101, 103
Basalla, G., 20–21, 30
Basic Education Certificate (BEF), 112, 116, 123
Bates, T. R., 101, 103
Bauer, M., 31

Bayly, C. A., 4, 30
Bechara, A., 132, 135
Becher, T., 38, 49, 179, 197
Bell, D., 6–7, 30
Ben-David, J., 179, 197
Benghabrit-Remaoun, N., 26, 114, 117, 119, 122, 123, 124
Bernal, J. D., 5, 30
Biglan, A., 38, 39, 49
Bingswanger, H. P., 165, 172
Bjarnason, S., 56, 77
Blackmore, J., 101, 103
Blaug, M., 6, 31
Bleiklie, I., 31
Bloom, D., 102, 103
Blume, S., 51
Blumenstyk, G., 92, 103
Boer, H. de, 15, 31
Bok, D., 31
Borges, J. L., 138, 141
Bougroum, M., 124
Bourdieu, P., 37, 49, 149, 152–53
Bowen, R., 102
Boyer, E. L., 194, 195, 197
brain drain, 95, 98, 102, 149, 187, 191
Brainard, J., 92, 103
Brazil, 145, 146, 158, 163
Brecht, B., 151, 153
Brhane, M., 105
Brint, S., 197
Bronfenbrenner, K., 102, 103
Brooks, A., 101, 103
Bryne, R., 102, 103
Buchbinder, H., 101, 103
Budu, J., 78
Burbules, N. C., 56, 77, 101, 104
Buxton, M., 41, 49

Callister, T. A., 101, 104
Canada, 6, 10, 21–22, 95, 101, 139, 145, 188
capitalism, 7, 10, 12, 81–84, 87, 100, 139, 143, 159, 161–62
Caplan, N., 43, 49
Cardenal, E., 153
Carnegie Foundation for the Advancement of Teaching, 193
Carpentier, A., 138

Carrera Damas, G., 173
Carvalho, J. M. de, 159, 172
Casanova, P. G., 146, 148, 150, 152, 153, 154
Castells, M., 31
"Cawthar" report, 134
Ceceña, E., 146
Cele, G., 102, 106
center-periphery, 21
centers of excellence (COE), 98, 187–88, 191
centers of learning (COL), 187, 188, 191
Chachage, S. L., 103
Chambers, D. W., 5, 21, 31
Charaffedine, F., 27, 129, 132, 133, 135
chartering and accreditation, 68, 176, 182
Chase-Dunn, C., 101, 104
Cherns, A. B., 50
Chile, 160, 193, 194
Chilundo, A., 78
Chomsky, N., 146
civilization/civilized, 140–44, 158–59, 171
Clark, B. R., 101, 104, 179, 183, 188–89, 190, 192–93, 197
Clark, W., 166, 172
cognitive structures, 42
Cohen, D., 39, 49
cold war, 8, 138, 171
Coldsteam, P., 56, 77
Coleman, J. S., 53, 69, 76, 77
collectivization of access, 86–87
Collins, H., 168, 172
Colombia, 160
colonial aggression, 143
colonial centers, 21
colonial elites, 22
colonial higher education, 21–22, 146–47
colonialism, 21–22, 131, 137, 141, 150, 158–61, 163
colonies/colonial peoples, 21, 139, 158
colonized regions, 5
commercialization of learning, 86–87, 91–92
commissioning research, 43, 62
commodification of knowledge, 10, 56, 86, 88
competitiveness, 1, 4, 5, 9, 14, 29, 53, 60, 97, 176, 187–88, 192
computerization of education, 86, 88
connectivity of institutions, 64, 66, 86, 89

Considine, M., 101, 105
Consultation Document, 57–58, 63
Cooksey, B., 105
Cordoba Reform, 147, 148
core values, 13, 88
corporate R&D labs, 2
corporatization of management, 86–87, 94, 175
corrosion of academic freedom, 86, 89–93
Council for the Development of Social Science Research in Africa (CODESRIA), 92, 102
Council of Science and Technology, 192
Court, D., 102, 104
credibility, 15–18, 36
Creole peoples, 142, 158
crisis, 30, 56, 72, 74, 81, 82, 108, 117, 122, 123, 128, 145, 175
Cronbach, L., 43, 49
Crosby, A. W., 4, 31
Cuba, 144, 150–51
CUDOS, 13, 18, 23, 191
Cueto, M., 161, 172
cultural diversity, 168
cultural studies, 91, 149

Dahl, R., 37, 49
Danner, H. E., 78
Darwin, C., 5, 21, 38
Davidson, C., 91, 104
De Ketele, J-M., 109, 124
de Léonardis, M., 115, 124
de Oliveira, F., 146
De Sousa Santos, B., 142, 154
de Weert, E., 101, 104
Delanty, G., 61, 62, 63, 78
democracy, 63, 64, 69, 82, 84, 108, 111, 126, 134
democratization, 53, 62, 99, 108–15, 148
democratization of knowledge, 62
Derrida, J., 151–52, 153, 154
Deustua, J., 160, 172
development, sustainable, 84, 164–68, 172
Dickson, N. M., 166, 172
Dill, D., 32
Diouf, M., 102, 104
disciplinary differentiation, 90
discrimination, 26–27, 46, 108, 110, 117, 129, 131, 133–34, 138, 158

Djeflat, A., 123
domain based research, 42
dominant classes, 140
dominant-dominated syndrome, 162
Drayton, R., 21, 31
Drucker, P. F., 5, 31
Durham, W. E., 165, 173

Ebrima, E., 94, 104
economic liberalization, 24, 55, 72, 74, 84, 152
economic participation, 27, 129
Ecuador, 140, 160
Edgerton, D., 31
education, access to, 9, 62, 72, 74, 86–90, 94, 97–99, 108–10, 111, 115–17, 133, 182, 184
"Education White Paper 3," 59
effects from economic and political pressures, 189–91
effects from future society, 185–86
effects from international perspective, 186–88
effects from past society, 184–85
effects from the nation state, 188–89
Egypt, 131, 133
Ehara, T., 183, 193, 197
Elena, A., 31
elites, 22–23, 40, 81, 94, 146, 169, 182, 184, 185
    colonial, 22
    political, 158–59, 163
Elliot, V., 102, 104
Elson, J., 139–40, 154
Eluard, P., 137
Elzinga, A., 44, 49
emancipation, 4, 158–59
employment, 26, 62, 93, 102, 108–109, 111, 113, 117–21, 123, 127
Enders, J., 101, 104
Enders, U., 196
engagement, 24–26, 53–77, 101, 127, 135
enrollments, 9, 18–20, 22, 24, 72, 74, 101, 112, 115–16, 125–26
epistemic communities, 35–37
epistemic drift, 44
Epstein, S., 168, 172
equity, 26, 59, 70–76, 86, 97–99

Ethiopia, 94
Etzkowitz, H., 1, 14, 31
Eurocentrism, 82
European, 82, 138–39, 141–45, 147, 159–62, 169, 171, 182–83
evaluation:
　connoisseurial, 47
　peer, 35, 37, 40, 47, 63, 182
Evans, G., 92, 104
external proletariat, 139

faculty development (FD), 194–95
female employment, 117–22
female faculty in African universities, 94–95
feminist paradigms, 94
Ferroukhi, D., 116, 124
fertility, 27, 129
Fieldhouse, D. K., 161–62, 172
finalization stage, 42, 46
first peoples, 143
Fisher, H. A. L., 40, 49
Fogg, P., 102, 104
Foresight Initiative, 41, 44
Foucault, M., 38, 49, 141
Franco, J., 137–38, 139, 154
Free Trade Area of the Americas (FTAA), 153
free-market economy, 87, 93
Fry, P., 78
Fulton, O., 196
Funtovicz, S., 31
Furtado, C., 169, 172

Gagarin, Y., 137
Gardner, H., 110, 124
Gariba, S., 78
Gaudet, J., 110, 124
gender balance, 74, 111, 115–22. *See also* Arab women
gender empowerment measure (GEM), 127–28, 132
gender gap, 94–95
General Agreement on Trade in Services (GATS), 57, 97, 100, 103
Germany, 82, 183, 188, 189, 193, 194
Ghana, 72, 74
Gibbons, M., 9, 15, 31, 32, 42, 48, 49, 50, 53, 78, 89, 101, 102, 104, 167, 172, 186, 191, 197

Gidley, J., 55, 58, 78
global knowledge society, 67, 78. *See also* knowledge society
global North, 80–82, 84, 90, 93–95, 102
global public good, 67. *See also* public good
global South, 80–81, 84, 85, 93, 94, 100
globalization, 1, 22, 26, 55–56, 79–88, 97, 100, 101, 108, 145, 152–53, 161–62, 175–76, 180, 181, 186–87, 192
globalization and higher education, 67, 84–89
Goldberg, D. T., 91, 104
Golding, D., 31
Gonzalez-Block, M., 49
Gootenberg, P., 161, 172
governmental power, modes of, 43
Grandin, K., 31, 32
Gregory Kohlstedt, S., 31
Griffin, A., 101, 103
Grossberg, L., 154
Grove, R., 21, 31
Gruzinski, S., 145, 154
Gumport, P., 189, 197

Haddab, M., 114, 124
Hakiki, F., 118, 120, 121, 124
Halperín, T., 159, 172
Hanney, M., 49
Hanney, S., 50
Hans Henrik, H., 103
Hawkesworth, M., 50
hegemony, 36, 50, 55–56, 79, 80, 83–84, 101, 139, 172
Henkel, M., 31, 43, 46, 49, 50
heterogeneity, 167, 171
higher education. *See* universities
Hirji, K. F., 102, 104
historical form of society, 12
Hodges, D. C., 101, 104
Home, R. W., 31
Houtart, F., 146
human capital, 6–7, 53, 101
humanities, 43, 90–91, 94, 102, 179, 195
Hunter, J. P., 91, 104
Huntington, S. P., 141, 154

illiteracy, 126, 129, 130
imperialism, 3–5, 82, 143, 149, 155
Inayatullah, S., 55, 59, 78

inclusion/exclusion, 165, 167, 169
independence, 39, 55, 107, 108, 111, 158
Indians, 145–46, 158
indigenous, 145–46, 158–61, 164, 169, 170
industrialism, 3, 5, 184
industry and commerce, 41, 45, 55, 58–62, 81, 83, 98, 99, 108, 119, 184, 192
inequality, 115–17, 129
information and communication technologies (ICTs), 43, 56, 66, 73, 85, 88–89, 91–92, 111, 185
interdisciplinary knowledge and research, 42, 91, 102
internal colonialism, 150
International Development Bank (IDB), 110
International financial institutions, 81, 82, 100
International Monetary Fund (IMF), 56, 97, 119, 177
Internet, 111, 145
Isaac, J., 37

Japan, 53, 82, 139, 175–79, 182–89, 192–94
Japan University Accreditation Association (JUAA), 176
Johnston, R., 50
Jones, G. A., 31, 196
Jongbloed, B., 32
Jordan, 126
Joseph, J., 101, 104
Juravich, T., 102, 103

Karseth, B., 31, 196
Kassimir, R., 78, 105
Kates, R. W., 165, 166, 173
Kellogg, A. P., 92, 104
Kenya, 72, 90, 102
Kerr, C., 43, 78, 182, 188, 197
Khaldoun, I., 136
Kiker, B. F., 6, 31
Kintzler, C., 108, 124
Kjaer, A. M., 16, 131
knowledge:
   as a social force, 5–6
   as source of social power, 1
   commodification of, 10
   common sense, 38, 39
   configurational, 41–42
   confrontation of, 160
   cultural hybridism, 160
   disciplined enquiry and, 43
   disciplines, dimensions of, 38–39
   distributed, 167–68
   hard and soft, 38–41, 43, 94, 143
   hermeneutics of, 37, 38, 46, 47, 48
   hybrid, 168–70
   hybridity of, 169, 170, 171
   hybridization of, 161
   institutional imperatives of science, 39
   internalist perspective of, 39, 40, 41, 49
   local, 97, 164, 167, 169, 170, 172
   non-scientific, 160
   obliteration of, 159–62
   obstacles to ideas of, 5
   ordinary, 38, 39
   other forms of, 139, 157, 161, 167
   paradigms of, 40, 42, 46, 61, 85, 91, 94, 99, 102
   positivist forms of, 36, 41, 46–48
   relation between university and, 177, 190
   restricted, 41–42
   role of, 2, 10, 13, 18
   specialization of, 35–37, 40, 42, 46, 91, 170, 177, 190
   spectrum of, 38–42
   subsumption of, 159–62
   traditional, 172, 175
   unrestricted, 41–42
knowledge based economy, 3, 9–11, 59, 87, 89
knowledge based society 1 (KBS1), 181, 185–87, 191, 195
knowledge based society 2 (KBS2), 181, 185–86, 191, 195–96
knowledge economy, 1–2, 5, 10–12, 18, 23–24, 27, 61, 66, 71, 75, 86, 97, 100, 156, 175–76, 186, 189
knowledge functions, 176, 178–81, 183, 184, 190, 191, 196
knowledge production, 42, 79–80, 83, 84, 85–86, 89–95, 96, 99, 102, 164, 168
knowledge society, 1–7, 10–12, 18, 23–27, 53–54, 60–64, 70, 73–74, 76, 80–84, 90, 97, 100, 101, 156, 175–76, 180, 183
Kogan, M., 24, 30, 31, 49, 50
Kohan, N., 147, 154
Krimsky, S., 31

Krohn, W., 50
Kuhn, T. S., 40, 50
Kumar, D., 4, 21, 32
Kuper, A., 50
Kwesiga, J. C., 94, 104
Kyvik, S., 50

labor market, 59, 100, 103, 128, 130
Lach, L., 14, 32
Lafuente, A., 31
Laing, D., 50
Lakhdar, B., 112, 124
Lakjaa, A., 117, 119, 123, 124
Land Grant Movement, 53
Lapointe, J-C., 110, 124
Laredo, P., 13, 32
Laroche, M., 6, 32
Larsen, I. M., 196
Latin America, 70, 71, 127, 128, 138–41, 146, 149, 158–61, 163, 169, 171
Latour, B., 167, 172
lay experts, 168
Leach, B., 101, 105
learned and artificial men, 142
Lears, J. T. J., 101, 104
Lebanon, 126, 129, 133
Lebeau, Y., 78
Lefort, R., 92, 104
Lem, W., 101, 105
Lescarret, O., 115, 124
Leslie, L., 4, 10, 32, 33
Levey, L., 78, 105
Leydesdorff, L., 31
licensing and patenting, 10, 14
Limoges, C., 31, 78, 172
Lindblom, C. E., 38, 39, 49, 50
lingua franca, 160
literary intelligentsia, 138
logic intrinsic to academic work, 190–91
López Pacheco, J., 153
López-Bassols, V., 101, 105
Lowen, R. S., 14, 32
Ludden, D., 101, 105
Lukes, S., 37, 50
Lustig, L., 101, 104

Mackinnon, A., 101, 103
MacLeod, R., 4, 21, 32

Magubane, Z., 101, 105
Mamdani, M., 102, 104
managerialism, 41, 47, 50, 51, 58, 60, 82
managerialist culture, 50
managerialist response, 50
Mangan, J. A., 23, 32
Manuel, F. E., 4, 32
Manuh, T., 72, 74, 78
Marcel, M., 32
Marginson, S., 101, 105
Mariátegui, J.C., 146, 154
Mario, M., 66, 74, 78
market, 2, 6, 8, 10, 11, 15, 17, 24, 25, 41, 44, 56, 62, 71–74, 77, 148, 160. *See also* free-market economy
  globalization of, 81, 166
  labor, 59, 98, 113, 118, 120
market failures, 50
market imperatives, 25, 97
market mechanism, 175, 176, 180, 185–86
market relevance, 66
marketization of the university, 87–100
Márquez, G. G., 138–39, 154
Martí, J., 139, 141–42, 144, 146–47, 150–51, 153, 154
Marx and Marxism, 38, 83, 151
Maslow, A., 6, 32
mass-consumption society, 7, 184
mass higher education, 6, 20, 182, 184–85
Matsuura, K., 150
Meek, V. L., 196
Meena, R., 94, 105
Menchú, R., 146
Mendelsohn, E., 50, 51
Merton, R. K., 5, 7, 10, 13, 32, 79, 84, 179, 187, 191
mestizo, 141, 159
metropolitan discourse, 138, 139
Mexico, 143, 145–48, 150, 160, 163, 171, 194
middle class, 87, 164, 169
Mignolo, W., 141, 154
Mincer, J., 6, 32
Ministry for Vocational Education and Employment (MFPE), 113
Ministry of Education, Culture, Sports and Technology (MEXT), 175, 187–88
mission-oriented research groups, 13, 17
Mitchell, T., 101, 105

Mittelman, J. H., 56, 78
Mkandawire, T., 82, 102, 105
Mkunde, D., 90, 105
mode 1 and mode 2, 42, 89–91, 185–86, 191
modern human being, 7
modern politics, 1
modernity, 164
modernizers, 140
Modes I and II, 42, 48, 89–91, 186, 191
Montgomery, E. B., 50
Morales, E., 146, 154
Morocco, 117, 133
Mozambique, 66, 72
Mulkay, M. J., 40, 50
Muller, J., 64, 78
multidimensional, 54, 65, 71, 75
Murphy, M., 102, 105
Musisi, N., 74, 90, 95, 105
Muwanga, N. K., 74, 95, 105

Nabudere, D. W., 81, 101, 105
Nagel, E., 38–39, 40, 50
Nandy, A., 164, 166, 173
narratives, 150, 156–59, 161, 163
Nassar, N., 127, 136
nation-state, 69, 71, 107, 176, 177, 180, 184, 188–89
National Institute for Academic Degree (NIAD), 176
national R&D systems, 169–70. *See also* research and development
native half-breed Creole, 142, 158
natural man, 142
Neave, G., 8, 32, 54, 61, 78
Nelson, C., 154
neoliberal, 56, 84, 96, 100, 102, 145, 148–49, 153
Neruda, P., 138
Netherlands, 193–94
new contract of science with the state, 9
new economy, the, 9
new International, 153
new internationalism, 152
New Zealand, 139
Newman, F., 67, 78
nexus between research, teaching, and learning, 176, 183, 195, 196
Ngome, C., 90, 105

Nobel Prize, 138, 146
non-governmental organizations (NGOs), 93, 94, 102
normative frame, 55, 75, 76–77
North, the, 151–53, 164
North American Free Trade Agreement (NAFTA), 145, 155
Nowotny, H., 31, 32, 35, 37, 50, 78, 197
Nyerere, J., 68

Office of National Statistics (ONS), 117–19, 120
Olea, V. F., 146
Olukoshi, A., 98, 103, 106
Oman, 133
Oppenheimer, J. R., 3, 32
Organization for Economic Co-operation and Development (OECD), 41, 86, 101, 177
Ortega, M. L., 31
Ortiz, F., 144, 155
Orwell, G., 61
Othman, N., 56, 78
Otieno, W., 90, 105
Ottoman Empire, 128
"our America," 138–39, 142, 144–47, 149
outcomes, 40, 45, 47, 48, 56, 60, 64, 193

Painter, M., 165, 173
pan-Africanism, 99, 103
Paraguay, 160
Parris, T. M., 165, 166, 173
Parsons, T., 50
partnerships, 62, 67, 87, 93, 192
Pavitt, K., 13, 32
Paz, O., 138
pedagogy of the oppressed, 150
people "transplanted," 139
people's power, 151
peripheral intellectuals, 159–60
Peru, 160, 161, 171
Petras, J., 80, 105
Philp, M., 36, 50
Pinch, T., 168, 172
Pires, M., 96, 105
Polanyi, M., 40, 50
political asymmetry, 164
political elites, 158–59, 163

politics, 48–49, 64, 75, 126, 129, 131–32, 134, 150, 152, 153, 160, 163, 164, 171
Popper, K. R., 40, 50
Porat, M. U., 12, 32
postcolonial countries, 22, 82
postcolonial patterns, 22, 91, 96
post-industrial society, 6, 83
post-modernism, 38, 40, 75, 83, 91, 94, 171
poverty, 69, 70, 74, 76, 109, 110, 140, 149, 164, 166
power/authority:
  communicativeness, 40–41, 49
  defining characteristics and models, 35–38
  internalist/intramural, 35, 36, 39, 40, 41, 49
  persuasiveness, 37, 38, 43
  within academe/secular, 36, 37, 45, 48, 97
  women and, 132–34
privatization, 68, 84, 86, 97, 99, 145, 148, 182
probing, 38
productivity:
  academic, 176–79, 187, 189, 191–96
  research, 71, 176, 177, 180, 191, 193, 195–96
  scientific productivity, 178–79, 195
  teaching, 176–77, 180, 192–95, 196
public engagement, 18
public engagement in science, 11
public good, 1, 55, 67–68, 75, 77, 84, 90
public sphere, 62, 149
public universities, 67, 73, 74, 86, 87, 90, 98, 102, 177

Qatar, 129, 133
quality, 13, 25–26
quality assurance, 68, 177, 180, 181, 189, 192–94

racism, 137
rationality, 43, 157
Ravetz, J., 9, 31
reform and revitalization initiatives, 72, 73–74
reformist discourses, 55

Reingold, N., 5, 32
religion, 84, 91, 98, 115, 128, 140, 143, 145, 153, 159
Republic of Korea, 193, 194
research:
  applied, 41, 42, 90, 94, 166
  higher education, 62, 98, 101, 179, 184
research and development, 101, 169–70
research and publishing, 89–90
research institutes, 2, 11, 16–17
research productivity, 71, 176, 177, 180, 191, 193, 195–96
Reyes, A., 143–44, 155
Ribeiro, D., 139, 140–41, 146–47, 155
Rifa'ha Rafe'e al Tahtawi, 125, 136
Rip, A., 7, 50
Roca, D., 147, 155
role of university, 177, 186, 190, 195
Romano, V., 153
Rongtan, H., 19, 33
Rosenberg, N., 14, 31, 32
Rostow, W. W., 7, 32
Rothenberg, M., 32
Ruggeri, G. C., 32

Sader, E., 146
Said, E., 137, 143, 155, 157, 161, 173
Sall, E., 73, 74, 78
Sall, H. N., 109, 124
Salmi, J., 66, 78
Saudi Arabia, 129
Sawyerr, A., 25, 32, 71, 73, 78
Saxenian, A., 14, 33
Schankerman, M., 14, 32
Schneider, A., 93, 106
schooling, 113
Schugurensky, D., 56, 78, 154
Schwarz, S., 50
Schwartzman, S., 31
science:
  development in, 162–64
  the good life and, 163–64
  institutional imperatives of, 39
  modern, 39, 161, 163–64, 171–72
  new contract with the state, 9
  power of the idea of, 157
  pure, 162, 167
  renewal of, 171–72

science and the state, 163
 social, 35–37, 41–44, 179, 195
 sociology of, 178–79, 187, 195
 technology for sustainable development and, 164–67, 168, 172
science agenda, 166, 167
science as a social or scientific field, 37
science policy, 1, 7, 10
scientific merit, criteria of, 39
scientific modern world-view, 157
scientific productivity, 178–79, 195
scientifically trained staff, 7
Scott, P., 31, 32, 45, 50, 67, 78, 197
Selznick, P., 33
Sen, A., 75, 76, 78, 164, 173
Shakespeare, 139
Sharabi, H., 134, 136
Shinbori, M., 179, 197
Singh, M., 102, 106
sit-ins, 148
Sklar, R. L., 78
Slaughter, S., 10, 33
Smallwood, S., 102, 106
Smith, J. G., 135, 136
social change, 118, 122, 138–39, 165, 175, 182–83, 186
social development, 57, 68, 71, 177, 180, 183, 189, 191, 195
social indicators, 76–77
social justice, 55, 60, 65, 73, 77, 110, 138, 147
social responsiveness, 58, 66, 69, 71, 73
social robustness, 35–36, 37, 40
social sciences, 35–37, 41–44, 179, 195
socially engaged, 53, 56, 74, 77
sociology of science, 178–79, 187, 195
solidarity, 148, 150, 166
Solow, R. M. A., 6, 33
Soludo, C. C., 92, 105
Sørensen, G., 103
South, the, 151–53
South Africa, 59, 94, 98, 102
Soviet Union, 138, 151
Spengler, O., 141
sponsorship, modes of relationships, 43–49, 111, 139, 193, 194
Sschwartzman, S., 197

stakeholders and stakeholders interests, 54–65, 71, 75–77, 87, 176
Steger, M., 105
Stehr, N., 101, 106
Storer, N., 38, 50, 197
structural adjustment programs, 69, 74, 81, 84, 93, 95–98, 100, 109, 118
Subotzky, G., 64, 78, 102, 106
suburban college-trained consumer, 7
Suppes, P., 43, 49
sustainable development, 163–67
Sweden, 194
Syrian Arab Republic, 133

Tanzania, 72, 90, 95, 102
teaching productivity, 176–77, 180, 192–95, 196
technological gatekeeper, 170
technology, 41, 45, 70, 83, 91–92, 161–62, 164–72, 184, 187, 195. *See also* information and communication technologies (ICTs)
technology transfer, 108, 111, 164, 170
Teferra, D., 102, 105, 106
Teichler, U., 50
Teilhard de Chardin, P., 153, 155
Teixiera, P., 32
teleological thinking, 12
teleology of modernity, 162
theology, 148, 150
third mission activity, 10, 15
Third World, 107, 108, 137–38, 150, 166
Thompson, K. W., 70, 78
Torres, C. A., 56, 77, 78
Toynbee, A., 139, 141
transculturation, 144
transnational corporations, 37, 42, 87, 92, 100
Trigo, A., 149–50, 155
Trist, E., 37–38, 42, 48, 50
Trow, M., 6, 7, 31, 33, 182, 184, 185, 197
trust, 13–18
Turk, J. L., 102, 106

Uganda, 72, 90, 92, 94, 95
underdeveloped countries, 147, 151, 169
underdeveloping countries, 151
unemployment, 70, 76, 118, 119, 159

United Kingdom, 44, 45, 47, 48–49, 111, 139, 193, 194
United Nations Development Programme (UNDP), 110, 125–27, 129, 132
United Nations Educational, Scientific and Cultural Organization (UNESCO), 97, 100, 108, 148, 150, 166, 177
United States, 45, 53, 80, 91–93, 95–96, 101, 102, 111, 138, 139, 141–42, 145, 148–52, 185, 191, 193, 194
Universal Basic Education (UBE), 113
universities:
  accountable, 8–9, 67
  African, 69–71, 73, 74, 86, 90, 94, 95–100, 101, 102, 103
  entrepreneurial, 10, 14–15, 72, 87
  entrepreneurial performance of, 14
  globalization and, 67, 94–99
  instrumentalist vision of, 10, 96
  legitimacy of, 15–16, 85
  marketization of, 87–100
  public, 67, 73, 74, 86, 87, 90, 98, 102, 177
  research, 13–14, 17, 23, 62, 97, 101, 179, 184
  role of, 69, 177, 187, 190, 195
  third mission activity of, 10, 15
  values of, 15–16
universities, individual:
  Cambridge University, 188
  Eduardo Mondlane University, 66
  Makerere University, 95
  National Autonomous University of Mexico, 148
  Oxford University, 188
  University of California-Berkeley, 102
  University of Córdoba, 147
  University of Dar-es-Salaam, 95
  University of Havana, 147
  University of Manchester, 45
  University of Oran, 122
  University of Toronto, 102
  Yale University, 147
university autonomy, 55, 87, 89, 97, 98
unskilled labor, 112, 129, 170
urbanization, 169, 184
utopian notions, 3, 4
utopian projects, 27, 138, 139

Van Damme, D., 67, 78
Van den Daele, W., 42, 50
Van Raan, M. J., 20, 32
Vaux, J., 50
Veney, C. R., 100, 102, 106
Venezuela, 160, 163, 171
Vessuri, H., 158, 170, 173
vocational education, 96, 100, 112, 113–14
Vogel, B. R., 78
Von Walden, 50
Vygostki, L., 110, 124

Walcott, D., 138
Wallerstein, I., 30, 33
Warnock Report, 50
Weingart, P., 41, 50, 51
Weiss, C. H., 46, 49, 51
Werquin, F., 124
Western civilization and culture, 20–24, 126, 132, 139–45, 151, 157, 162, 169, 188–89
Whitley, R. D., 35, 41–42, 50, 51
Widmalm, S., 31, 32
Wilde, O., 139
Wilson, R., 102, 106
Wirt, F., 51
Wittrock, B., 49
women and authority, 132–34
Woodhall, M., 101, 106
work organizations, reputational, 35–36
World Bank (WB), 57, 73, 96–97, 100, 102, 109, 177
World Trade Organization (WTO), 56, 57, 73, 97, 108, 177
World War II, 53, 125, 188
Wormbs, N., 31, 32

Yemen, 129
youth, 87, 113–15, 118, 119, 123, 138

Zapatista Army of National Liberation Army (EZLN), 145
Zeleza, P. T., 96, 98, 101, 102, 103, 105, 106
zero, discovery of the, 143
Ziman, J., 9, 33
Zuoxu, X., 19, 33

Made in the USA
Monee, IL
03 May 2026

49437442R00122